AF876275

CHAOS & The Missing BUTTERFLY

The Mystery Hunters Series

Book Two: Mathematics

CHAOS
& THE MISSING
BUTTERFLY

KURTIS C.R. PALMER &
RAMONA R. PALMER

WOMB TO BLOOM TO CLASSROOM LLC / NEW YORK

Copyright 2019 Womb to BLOOM to Classroom LLC, by Kurtis C. R. Palmer & Ramona Palmer.

All rights reserved. No part of this publication may be reproduced, distributed, or transmitted in any form or by any means, including photocopying, recording, or other electronic or mechanical methods, without the prior written permission of the publisher, except in the case of brief quotations embodied in critical reviews and certain other noncommercial uses permitted by copyright law. For permission requests, write to the publisher, addressed "Attention: Permissions Coordinator," at the address below.

Womb to BLOOM to Classroom LLC, Kurtis C R Palmer & Ramona R. Palmer
P.O. Box 1084, Canandaigua, New York 14424

1st Edition Printing March 2021

www.W2BLOOM2C.com or www.SmarterBrighterKids.com

Publisher: Womb to BLOOM to Classroom LLC, Canandaigua, NY 14424

Publisher's Note: This is a work of fiction. Names, characters, places, and incidents are a product of the author's imagination. Locales and public names are sometimes used for atmospheric purposes. Any resemblance to actual people, living or dead, or to businesses, companies, events, institutions, or locales is completely coincidental.

Book cover illustration by Rey Brava

Photos purchased from IStock.com & Shutterstock.com

Copyright permissions received and granted.

Ordering Information: Special discounts are available on quantity purchases by corporations, associations, and others. For details, contact the publisher at the address above.

ISBN 978-1-953509-03-1 Softcover Edition

Printed in the United States of America

Dear Reader,

Thank you for taking this journey. You are an important participant in this story. *You and a friend become part of the story by filling in your names in the blanks, as Sarah and Christian's best friends.* We have **7 questions in bold lettering** in each of the 7 chapters for you to answer. Enjoy the mystery hunt! Remember the famous quote from Shakespeare's play *Romeo & Juliet*, **"What's in a name?"**

❖ ❖ ❖

There are two blank pages in the book.

Sketch a picture of you and your friend in one of the book scenes while listening to your favorite song.

***If you're drawing this in the softcover book, use a pencil first.* When you're finished, color in your drawing. *Sign your name* to the drawing and enter the name of your favorite song and the musical artist.**

We dedicate this book to our beloved
Betty (Toney) Ross.
Our Tata, grandmother, lived by the Golden Rule,
Matthew 7:12 **-** *So whatever you wish that others would do to you, do also to them, for this is the Law and the Prophets.*
She is generous, tireless, wise, gentle, funny, sweeter than sugar and honey, and most of all, LOVING!!!

CONTENTS

CHAPTER ONE

ANOTHER SUMMER ANOTHER MYSTERY

Sarah and Christian both dream mysterious dreams. These dreams transport them around the corner, around the globe, and through the universe and back. They're first cousins and best friends who plan *mystery hunts* every summer.

Christian's and Sarah's grandmother, Tata, had this same gift of prophetic dreams. She called them 'waking dreams' because she dreamt of echoes from the future and then, suddenly, woke up.

She instructed her grandchildren to be cautious of the invisible force that would try to disrupt or stop their quests. She warned them, "This invisible force will try to frighten you, change your circumstances, alter your path, and place obstacles in your way. You were given a special gift. Use it wisely."

As Christian and Sarah walked to the library, Sarah said, "Tata was right. There seemed to be an invisible force trying to stop us from finding Professor Septenary and even trying to stop him from coming back home. I think we should focus on solving these mysteries as quickly as we can, so no one gets hurt."

Christian said, "But these could just be things that happen every day. I asked Tata what the invisible force was, and she just smiled and said, 'You each have to discover the answer for yourself. Just be careful.' But if she's right, I agree that we have to solve these mysteries as quickly as possible before anyone's seriously hurt. Well, I'm ready for a new mystery to solve."

"I might have something," Sarah said. "I read an article online about a lady who got robbed at home. The suspect took a butterfly."

Christian half-smiled and asked, "What's so important about a single butterfly?"

Sarah said, "I'm not sure, but the online article mentioned a mathematical idea called *Chaos Theory*."

Christian asked, "What does math have to do with butterflies?"

Sarah said, "*Chaos Theory* or *The Butterfly Effect* is, I guess, a really important theory."

Christian asked, "What else did you find out?"

Sarah said, "The woman is Ms. Parvana, and when her grandparents were living in Connecticut, they were friends with Professor Edward Lorenz, the meteorologist, who was also a mathematician, studying *Chaos Theory*. I'll see if she can meet with us."

Christian said, "Sounds good. We should meet with her alone. If she's interested, we can call _______________ and _____________, and ask them to join us later."

"Great idea!" Sarah said. "I'll call her this afternoon."

◆ ◆ ◆

The next morning Sarah called Ms. Parvana and offered to meet with her.

Ms. Parvana said, "Yes. I saw your names in the newspaper when you were helping in the search for Professor Septenary. I'd love to meet with the two of you. Please stop by this Saturday at noon. I'll make lunch."

Sarah said, "Thank you. We'll be there."

After the phone call, Sarah reached out to Christian. "I spoke with Ms. Parvana. She offered to make us lunch at noon. She read the newspaper article on our search for Professor S."

"We're getting a good rep'," Christian said. "I'll come by your house at 11:30, and we can walk to her house together."

"Sounds great!"

◆ ◆ ◆

On Saturday, Christian knocked on Sarah's door. "Ready to take a walk?"

"Yeah, I want to bring my laptop to take notes."

Christian rang the doorbell at Ms. Parvana's home.

Ms. Parvana opened the door. "Welcome, it's so nice to meet both of you. Please come in, and we'll enjoy lunch and share stories."

Sarah said, "This is a great house! Have you lived here long?"

"My parents owned this home, and I purchased it from them several years ago. Please come out to the patio."

Sarah and Christian followed Ms. Parvana to a table set with red, blue, yellow and purple wildflowers, butterfly plates and glasses, a jar of sun tea with mint leaves and fresh lemon, and a large plate of fruit.

Sarah said, "The flowers are so beautiful. I love your plates and glasses. The food looks super good."

"Thank you. Please sit and let me serve you some iced tea and lunch. I've always loved butterflies, and I'll tell you the story of my framed butterfly."

The three enjoyed a delicious lunch. Ms. Parvana baked fresh sugar cookies in the shape of butterflies. The cookies, decorated with colorful sugar sprinkles in geometric designs, melted in their mouths.

Christian said, "We read about your butterfly and Professor Lorenz."

Ms. Parvana said, "Professor Lorenz is the meteorologist who wrote his theory on *The Butterfly Effect* or *Chaos Theory*. The Professor was a close friend of my grandparents. He gifted my grandparents a beautiful butterfly in a picture frame. I've loved butterflies since early childhood, and I would look at the framed butterfly every night before I fell asleep. They're so elegant and colorful, and as a child, I always dreamed of flying. My parents explained Professor Lorenz's theory, but I didn't research *The Butterfly Effect* until I was older."

Sarah said, "Christian and I dream of flying! My favorite pillow is decorated with butterflies. I dream when I fall asleep on my pillow."

"Butterflies are a symphony of colors, each unique, floating, and dancing mid-air. I teach special needs children at our local elementary school. I decorate my classroom each year with numbers and butterflies. This past school year, I taught seven second-grade students, all of whom have autism. Two of my students, Vanessa Nova and Mary Cartwright, excel at math and both girls love butterflies as much as I do."

Christian asked, "What's autism?"

Ms. Parvana explained, "Children with autism have difficulties speaking and communicating, and repeat behaviors. But many of these children are excellent at math, and possess amazing memories. My students are kind to everyone they meet. Each time I look at their young, delicate faces and listen to their thoughts and dreams, I'm reminded of butterflies. I envision each child floating and dancing mid-air, spreading their wings to fly higher and higher. One of the reasons I'm so dedicated to teaching is that I believe in amazing possibilities."

Sarah said, "Wow! That's so nice."

Christian smiled and asked, "Why was this butterfly so special to Professor Lorenz?"

Ms. Parvana said, "Professor Lorenz used the example of a ___________ when first describing *Chaos Theory*. Later, a butterfly was used to explain this phenomenon, and the same day the Professor heard his theory called *The Butterfly Effect*, he went outside his home, and a butterfly landed on his hand. He kept this butterfly and then framed it, to remind him of his theory."

Christian asked, "Is it worth something?"

Ms. Parvana answered, "It's not worth a great deal of money, but it's sentimental to me. A reporter interviewed me a few weeks ago about my family heirloom."

Sarah said, "It makes sense. The person who took the framed butterfly probably read the newspaper article. They may love mathematics, or butterflies, or both."

Christian asked, "Do you have a photograph of the framed butterfly?"

"Yes. The newspaper reporter sent me a lovely photograph of the article with the framed butterfly. You may take it with you. I also have photographs of my grandparents standing next to our family heirloom."

Christian said, "If you don't mind, I'll take a photo of the article with my cellphone. I don't want to lose the article. It's one of your last photos of the butterfly."

"Thank you. The framed butterfly is my prized possession. I think of my grandparents every time I look at it. I use it to explain *The Butterfly Effect* to my second-grade class every year."

Sarah added, "I want to make sure we receive your permission to investigate this theft, and your permission to invite two of our friends to join us."

Ms. Parvana said, "I don't want anyone to be in danger. I reported the incident to Detective Dawson at the police department."

"He's my uncle," Christian said. "I keep him up to date on everything we're investigating. Safety is important to us, too. Don't worry."

Ms. Parvana said, "Thank you."

Sarah asked, "Did the police share any clues with you?"

"Yes. I don't think the police found fingerprints, but they discovered a shoeprint outside my window and took a photo of the print."

Christian asked, "Did the police say anything about the shoeprint, like its size or sole design? Is anything else missing?"

"No. They didn't give me any information on the shoeprint. I checked my entire home, and nothing else is missing. Detective Dawson asked me the same question."

Sarah said, "Thank you for lunch. It was a nice afternoon. We're so excited about helping you. We'll let you know if we find any new clues."

Ms. Parvana said, "Thank you."

Sarah and Christian walked home.

◆◆◆

Later in the evening, Sarah fell asleep on her butterfly pillow.

Sarah dreamt of seven butterflies floating in the sky, but they only flew so high. She saw a precious little girl, twirling around her bedroom wearing a butterfly dress and butterfly sneakers. Sarah asked the little girl her name, and the sweet girl put her two fingers up to her pink lips and whispered, "Butterfly." The little girl wore a sparkling butterfly hair clip holding up her curly, long brown hair. The butterfly on her hair clip reminded Sarah of the photo of the framed butterfly. There were three books on a shelf in her room. One book had butterflies on the cover. Another book had the word "Relative," *written in calligraphy lettering. One book had gold embossed lettering with the word* "Symmetry." *Sarah grabbed the book on butterflies, but when she looked down, she saw the title of the book was* "Relative."

Sarah suddenly woke up as her head dropped to the bed. She felt as though someone pulled the pillow out from under her head. Her butterfly pillow was across the room. She whispered to herself, "How did my pillow get way over there?" She shook her head and said, "Weird, but what an amazing dream! I wonder what it means."

The next day Christian and Sarah met at Sarah's home and called their two friends, ____________ and ____________. Sarah explained the next mystery hunt, and they were both excited for a new adventure.

______________ said, "I'm on my way to your house. See you soon."

Both friends arrived at Sarah's home. Sarah invited them in, and they sat at the dining room table. Sarah opened her laptop and said, "I took notes when Christian and I met with Ms. Parvana. Both Christian and I think the person who took her family heirloom read the newspaper article about the framed butterfly."

_____________ asked, "Does she have any idea who took it?"

Christian answered, "No, but she agrees someone must have read the article. She doesn't lock her door because we live in such a safe town, but I'm sure she'll lock it now."

_____________ said, "Sarah, you love butterflies. You must be thrilled!"

"Yes," Sarah said. "Christian and I agree whoever took the butterfly is more interested in its math connection to *The Butterfly Effect* or *Chaos Theory*. Ms. Parvana believes it may be worth something, but only because of its connection with Professor Lorenz. It's not the value of the butterfly. This family heirloom is precious to her because it was a gift from her grandparents."

____________ said, "I can understand why she feels that way."

Sarah asked, "Christian, will you look up both theories on my laptop?"

Christian began reading, "*Chaos Theory* is a branch of mathematics focusing on the behavior of systems sensitive to certain conditions. *The Butterfly Effect* describes how a small change in one simple system can cause enormous differences in a later or larger system. A metaphor, or a symbol, for this behavior, would be a butterfly flapping its wings in Japan, causing a hurricane in Florida."

___________ said, "I remember an example of this in math class. Professor Lorenz was a meteorologist.

"Our teacher explained his *Chaos Theory* with a history story. My teacher, Ms. Flutura, used the example of Rosa Parks. You'll think this is amazing. **Ms. Flutura's name means _____________.**

"Ms. Parks was a black woman who refused to give up her seat on a bus to a white man. There was a law called segregation in our country. People of color couldn't sit near the front of a bus or drink from 'White' drinking fountains. **Ms. Parks was _______________, or taken into custody, and many boycotted or refused to ride the bus. This event grew into the _________ ________ Movement.** They passed laws against segregation and discrimination toward people of color. So, the example of the butterfly causing a hurricane reminds us Rosa Parks was like the butterfly. My teacher said, 'but instead of a destructive hurricane, her small act changed history.'"

Sarah said, "We can agree the missing butterfly is important to Ms. Parvana for personal reasons. It's also important in history as a symbol of a famous math theory."

______________ asked, "So, where do we start?"

Christian answered, "We need to find out if anything like the butterfly has gone missing. We can search online and at the library for a pattern. If there are no similar crimes, we can focus on our original theory. Someone read the article on the framed butterfly, and they took it to sell."

Sarah said, "I don't think the person took it to sell. Nothing else was taken from Ms. Parvana's home. The person just wanted the butterfly."

______________ said, "Or maybe the person heard sounds outside when he was in her home, and it spooked him. Then he took off, and there wasn't time to steal anything else. Anything's possible. We can meet at the library tomorrow at 10:00. I'm walking my little cousin to the library to choose a few books and watch a puppet show."

______________ said, "Sounds great. See you tomorrow."

❖ ❖ ❖

The next morning the friends sat at a large table at the library while ______________'s little cousin enjoyed the puppet show.

Christian said, "I checked a few things on my computer last night. There was chatter on the article about the butterfly, and two of the comments were a little suspect. One person at iambutterfly777td@gmail.com commented on its beauty, and thought the framed butterfly would be a nice addition to her collection. Another writer wrote butterflies should be free. But I have no idea where they live. It could be anywhere in the world."

______________ said, "Yeah, but it doesn't rule them out, they might live around the corner. Or they hired someone to steal the butterfly if they wanted it badly enough."

______________ said, "Or they have nothing to do with it, or they took it for a different reason. Again, anything is possible."

______________ offered, "I'll email those who commented on the theft and keep an eye out for other online posts about the butterfly."

Sarah said, "It's a first step. I'll research online and look to see if there are other missing butterflies. I need a few days. We can meet at my house for lunch on Thursday. Tata always says, 'When we break bread, we share an act of generosity of spirit ... an act of unity. We move one step closer to each other.'"

Christian said, "We love Tata! Lunch sounds amazing! The rest of us will research information on *The Butterfly Effect* and *Chaos Theory*."

◆ ◆ ◆

Three days went by, and the friends worked on the mystery of the missing butterfly.

They met on Thursday for lunch and shared their research.

"I researched the reason we call a butterfly, a butterfly," ____________ said. "The word 'butterfloege' means butterfly because the yellow brimstone butterfly has a 'butter' yellow color. In other languages, butterfly means 'milk thief'; because we make butter from milk. **In the Greek culture, the butterfly represents the ___________ or immortality.** Some people believe they turn into a butterfly to fly to heaven."

_____________ said, "Cool! I found more information on *The Butterfly Effect*. **Professor Lorenz worked at ___ ___ ___.**"

Sarah said, "This reminds me of our last adventure. Professor Septenary's decision to search for the meaning of the number seven created a worldwide search for him. People around the world read his online journal. We learned facts about the mysteries of the number seven, our seven continents, and cultures and religions. One small act turned into a much larger event."

_______________ said, "I researched information on *Chaos Theory*. Chaos is an event we can't predict because of moving and changing conditions."

Christian suggested, "The four of us should go online to see if the butterfly is worth anything."

______________ said, "I looked it up. The framed butterfly might be worth over $1,500."

Sarah said, "I still don't think they took it to sell. There wasn't anything online about other stolen butterfly objects."

Christian said, "I found something online. Someone stole two books from a home in Livingston County. The book collector had the second edition of Euclid's book,

Elements, and *Liber Abaci* by Leonardo of Pisa. Livingston County is right next to ours. The owner's name is Mr. Bigollo. I'll call him to see if he'll meet with us."

____________ added, "We'll help find his missing books."

Sarah asked, "What else did you read in the article?"

Christian said, "It gave some background on the two mathematicians and their theories. **One is *The Golden Ratio*, and one is the *Fibonacci* ________.** There's a strange connection between the two theories ... they both produce the same number, 1.618. If the same person took the three objects, someone's interested in collecting math books and objects."

____________ said, "I'll check other online sites when I'm home tonight. Amazon, eBay, art, and collector's sites list stuff for sale. It'll take a few days."

____________ said, "If just the two of you meet with Mr. Bigollo, ask him if he locked his door. Also, ask him if anything else was missing and does he have any idea who had access to his home. Also, please ask him if the books are worth anything."

Christian smiled. "We always bring a list of questions, but you learned a lot about important questions from our first mystery hunt. Nice job!"

____________ said, "Well, it begins."

Everyone smiled.

◆ ◆ ◆

The next morning, Christian called Mr. Bigollo.

"Good morning, Alex Bigollo speaking."

"Good morning, Mr. Bigollo. My name is Christian. I read an article in the newspaper and online about the theft of two of your books. My cousin Sarah and I, and a few friends, take on mystery hunts every summer. We might sound young, but you can contact Detective Dawson, my uncle. We assisted the police in a past investigation."

"I remember reading an article in the newspaper about four students who were searching for a lost mathematician. You must be them."

"Yes, Sir. We're the four mystery hunters," Christian said.

"We appreciate any help. I'll meet with you. The four of you are too young to drive," Mr. Bigollo chuckled.

Christian suggested, "We can meet at our library in town at 1:00."

He provided Mr. Bigollo with the address. Christian looked up Mr. Bigollo's Facebook page, so he would recognize him at the library. Christian called his three friends. They were excited to meet Mr. Bigollo. They each had questions.

FRAMED BUTTERFLY

CHAPTER TWO

PHI, PI, OR PIE?

Did Mr. Bigollo hold a clue to solve this mystery? Did he know more than he was telling Christian? Was Mr. Bigollo connected to the missing butterfly? The four friends walked to the library to meet with Mr. Bigollo. Christian was sure he'd recognize him from his FB photo.

The library was bustling with children sitting at pint-sized tables with parents reading to their children, puppet shows, painting, and coloring games going on. The friends saw a tall thin man, with dark curly hair and round tortoise-shell reading glasses, sitting at their favorite table.

Christian recognized Mr. Bigollo, walked over to the table, introduced himself, and said, "Mr. Bigollo, this is my cousin Sarah, and our two friends, ______________ and ________________."

Mr. Bigollo stood up and shook the four friend's hands. "It's wonderful to meet you. Thank you for offering to help in the search for my books. I planned on arriving early this morning to do some reading, but I had trouble starting my car. Then, just as it started, a branch fell from a tree in our yard. It's a huge branch, and if my car hadn't started, it would have crashed right through my front window. It was a close one."

Sarah asked, "Are you okay?"

"Yes, thank you. It was very strange. I inspected the branch, and it seemed to have been ripped from the tree. I just had my car tuned and inspected, so I'm not sure why it didn't start at first."

Sarah and Christian looked at each other.

Christian said, "We're glad you're okay. Would you tell us about your books?"

"Yes, thank you for asking. I became a professor of literature because my grandfather introduced me to books and encouraged me to read at a very young age. I enjoy sharing this story. My grandfather had a profound passion for mathematics and reading. Early in my childhood, when I visited his home, he'd read me books on math, science, and history. I spent weekends at my grandparents' farm, helping out. My grandfather had a small room filled with hundreds of books. I remember books on the floor, on bookshelves, on his desk, in boxes, and stacked up to the beamed ceiling. There were leather hardcover books and a few books embossed with gold or silver. He filled his library, for lack of a better term, with math books, brilliant novels, biographies, books on science and astronomy, and children's books. He treasured each book.

"I can still hear him saying, 'I dream of time travel. Every time I read, I travel to the past, and I'm in the present. I travel to the future when reading science fiction. Authors, many of whom lived hundreds and even thousands of years ago, wrote their thoughts, dreams, and imaginings. These books are their introduction to our world. It's as if each author is sitting here in our library, listening to us read these great works. Think of all the eyes that have read these words over the centuries. Think of the lips that whispered the words from each book. Think of how these words then moved the reader to place their feet on a different path ... search for a new direction. Books can shape a life, enrich our understanding, encourage a new beginning, lift the spirit, spur the imagination ... books can change a life.'

"After working twelve hours a day on his farm, and after my grandmother's delicious meals, we would enter his library. He asked me to close my eyes and spin around once and point to any book. He read a few pages from each book, and we continued reading that book every weekend until we reached the last page. While there was much I didn't understand, he allowed me to ask as many questions as I liked. **The first book I pointed to was Euclid's, *Elements,* and the second book I**

pointed to was *Liber Abaci,* by Leonardo of Pisa, known as ____________. We read over 100 books together.

"I'd like to share his story with you about the first time he decided to open The Bible. He told me that decision changed the 'essence of who he was and who he would become.' I wrote down what he said and keep it close to my heart."

Mr. Bigollo pulled a wallet from his jacket pocket and unfolded a small piece of paper. "My grandfather wrote, 'All along, The Truth, the answers to my questions about life, had been sitting on a bookshelf in my home long before my birth. This Book, with its frayed edges, stood silently during my childhood and formative years and into adulthood. It made no sound. It stood guard over our home. It was the last Book I opened for the answers. Yet, The Truth had been sitting there, waiting for me to pull it from the shelf and open it to the first page. *In the beginning*....' I still have his Bible, a family heirloom, and I still love reading because of my grandfather."

Sarah said, "Your grandparents sound wonderful. It would take three times as long if each of you read a book alone. I really love the story about the Bible."

The friends were mesmerized by Mr. Bigollo's stories.

"Reading a book with your family is an exploration. Many times, my grandmother came into the library and sat in her favorite rocking chair. My grandfather built this magnificent oak and walnut chair by hand as an anniversary present for my grandmother. He carved their initials and marriage date into the wood on the back of the chair. She listened as we read and then she fell asleep. Every time her eyes closed, my grandfather covered her with a soft, light blue fleece blanket, and we continued to read. We would discuss everything we read. I asked him so many questions.

"My grandfather gifted me the rocking chair and his books. I have so many wonderful memories. Every Thanksgiving, I filled a bowl with my grandmother's homemade cranberry salad. This delicious creation had fresh ground cranberries with orange peels and walnuts, mixed in raspberry JELL-O. I took my bowl to my room at their home, turned on the small stained-glass lamp, and read, read, and read until I fell asleep. My grandmother gifted this recipe to my wife, and she makes it every year at Thanksgiving. After Thanksgiving dinner, I still take a small bowl of this magical fruit elixir to my bedroom and read, read, and read until I fall asleep. Thank you for listening to my stories. It probably doesn't surprise you that I teach literature."

The friends smiled, and Sarah whispered, "Cool!"

"I plan to give a portion of my collection to our nephew, who lives in Great Britain. The rest of the collection is willed to my wife's niece. My wife and I don't have children. Our nephew worked in France for a few years at a large bookstore. He now owns an antique bookshop on Church Street. We've been estranged for a few years. He asked me to give him two books for his shop. Come to think of it; both books are my missing books. I never want these books sold for profit. I informed him I'd donate them to a museum before selling them. He said he understood, but he never returned my phone calls and emails after our talk. He has no idea we've willed some of our collection to him."

Christian asked, "I'm sorry to bring up this possibility, but do you think your nephew could have something to do with your missing books?"

Mr. Bigollo quickly replied, "Absolutely not! We may have our disagreements, and he might resent me for not giving him the books to sell, but he'd never steal them or put us in harm's way. I didn't even mention this to the police. There was a time when he was hanging around with the wrong crowd, but that was several years ago. Now he's an upstanding citizen with his own business."

Sarah said, "I'm sure you're right. He must love books if he owns a bookstore. He'll be happy when he finds out you left him some of your family heirlooms."

_____________ asked, "Are the books worth anything?"

"Yes," Mr. Bigollo answered. "Euclid's book is a second edition that is priceless. *Liber Abaci* is also a second edition, worth over $30,000. My grandfather used gloves to turn each book's pages, and I do the same thing. I insured both books, but I want them returned. I hope whoever has the books doesn't destroy or damage them. What's most valuable to me is the memory of my grandfather and my time with him, reading our first two books together."

_____________ said, "I don't think the suspect plans on damaging the books. We're working on another case of a missing object. The person taking these objects must love math. Are you missing anything else from your home?"

"No," Mr. Bigollo answered, "and there were other valuable items in my home. I filed an insurance claim, but I just want my books returned intact."

Sarah said, "Mr. Bigollo, the other missing object is a framed butterfly, once owned by Professor Lorenz. The suspect only took the butterfly."

"I'm familiar with *Chaos Theory* or *The Butterfly Effect* and Professor Lorenz. You should also know I contacted the police."

Christian said, "My Uncle Raymond Dawson is a police detective, and our group works with him. We'll meet with him and tell him our discoveries."

"Yes, I spoke with Detective Dawson at the police station. Thank you. I appreciate your help. Both cases might be connected."

____________ asked, "Could you explain these theories?"

Mr. Bigollo replied, "They're both intricate theories, even though the mathematicians who proposed these theories lived centuries apart. **Euclid's *The Golden Ratio* theory is in the sixth chapter of his book, *Elements,* represented by the __________ letter Phi (φ). It has a ratio of ________**. Pi has a ratio of 3.14. His sequence is a line continually moving toward Phi but goes on forever. I remember reading the following quote, 'Mathematicians have heralded *The Golden Ratio* as the most beautiful ratio in art and architecture.'

"We find *Fibonacci's sequence* in nature and art. Fibonacci proposed a sequence of numbers in which the next number in the sequence is found by adding the two previous numbers in the sequence, 0 + 1 =1, 1 + 1 =2, 2+ 1 =3, 2 + 3 =5, etc. **This sequence of numbers creates a __________.** We can see this in sunflowers, seashells, and even galaxies. Thank you again. I need to leave for home now, but I suggest the four of you read a few books on both theories. You'll discover these numbers in art and nature, and even in the human face."

Christian said, "Thank you so much! We'll stay here and get right on it. I'll tell my Uncle Raymond, and we'll update you on anything connected to your books."

◆ ◆ ◆

The four friends searched for books on *The Golden Ratio*, and the *Fibonacci sequence*. Each book included beautiful photographs of art, flowers, and plants, where these designs manifest. The friends walked home after staying for another hour at the library.

The next day they met at Sarah's home for blueberry pie and Christian's favorite, vanilla ice cream. As they were talking about their adventure, Sarah said, "This delicious pie reminded me of a fun fact. **As I was researching Pi, I discovered National Pi Day is ___/___.**"

____________ said, "I read the same thing! It's awesome!"

Christian said, "I called my Uncle Raymond. He's the lead detective on the case of the missing framed butterfly. We both agreed the two cases could be connected. He's working with the lead detective in Livingston County. They both agreed my uncle would be the lead detective on any missing math objects. Uncle Raymond has a new boss that's pretty tough on crime."

______________ said, "Does he get along with his new boss?"

Christian answered, "Yes, and no. My uncle is tough, but he has a kind heart. He tries to help some of the suspects. He believes most people can change if someone cares and makes an effort. I overheard my uncle talking to my dad. My uncle's new is boss doesn't think anyone can be helped."

______________ said, "I'm with your uncle."

The friends nodded in agreement.

Sarah said, "I researched online sales of math objects. Unless the sales are on sites I can't access, no one is selling the framed butterfly or the books."

Christian added, "The police have an entire department dedicated to artifact and antiquity thefts. My uncle described their computer division. Their computer system moves at lightning speed. The police are checking online for the sale of these items. He reminded me there are websites none of us should search."

______________ said, "I agree. There's so much for us to do, but we'll let the police do their job."

Christian said, "My uncle will contact me if they find anything. Of course, there's information he can't share with us."

______________ asked, "What's next?"

"We should search online for other missing math objects and research math theories," Sarah said.

Christian said, "Yeah. Why don't we meet with my Uncle Raymond? I'll invite him over to my house."

______________ said, "Great idea!"

◆ ◆ ◆

The next day, they knocked on Christian's door. Uncle Raymond was sitting at the dining room table. "Hello, my junior detectives," Detective Dawson said with a smile.

Christian announced, "I told my mom we were studying *Phi*, *Pi*, and the *Golden Ratio*. In honor of our search, she made chicken pot pie soup with a golden biscuit, a pizza pie, and a fruit pie. She cut the pies into pieces or ratios and carved the *Phi* and *Pi* symbols into the golden butter crusts. I updated Uncle Raymond on both cases."

Everyone smiled.

___________ said, "It looks great!"

"Thanks again for agreeing to help," Uncle Raymond said. "You worked day and night to search for Professor Septenary. I'm taking the lead over this investigation. The four of you must stay safe. Please contact me with any information, and we'll follow up on your leads. There's no evidence of the framed butterfly or the books being sold. It might mean the suspect took these family heirlooms to keep for himself or sell to a private collector. We'll notify the FBI if we discover other items are missing outside of New York State. They're responsible for interstate searches when more than one state is involved. **I'm sure you know this, but just in case, FBI stands for __________ _______________ of _______________."**

"Thanks," Christian said. "We understand you can't provide information on the case. We thought we should focus on other reasons these objects went missing. Other than selling the objects, the only other reason is to keep them."

"I agree," Uncle Raymond said.

Sarah said, "Christian will update you on everything we find."

Uncle Raymond said, "Thanks again. Christian, thank your mom for a great lunch. She started cooking as a young girl, and her practicing made her cooking and baking perfect. Well, I need to return to work. Remember, stay safe."

The friends called it a day to spend time with family.

◆ ◆ ◆

Sarah and Christian attended a family picnic at their cousin's house. Alex began racing cars when he was very young. Now, at age 13, he's winning races against many older drivers. Kennedy has so many friends, keeping her busy every weekend, camping, swimming, and having sleepovers.

The family was celebrating Alex's latest win. Toni, Alex, and Kennedy's mom ordered a huge congratulations cake, decorated with red, white, and blue race cars. Everyone enjoyed pizza, chicken wings, and sub sandwiches. Christian was thrilled! There was one layer of just-plain-vanilla cake.

Aunt Toni decorated the patio with race car balloons. Alex and Kennedy drove their cousins around the yard on green and red go-karts. Their dad watched the kids on the go-karts, making sure everyone was safe. There were pony rides for the younger kids. A specialty company delivered a bounce house and a rock wall. A bunch of the parents jumped up and down in the bounce house with their kids. Everyone questioned Christian and Sarah about their mystery hunts.

Cousin Henry, who was seven, had so many questions. "Are you on another mystery hunt? Do they pay you? Where are you going next? Do you go searching for gold and diamonds? Do you get to ride horses? Do you fly planes? Do you get to fly kites? Are you going to be on television again? Can I go with you sometime? What's your favorite dinosaur? Do you like chocolate or vanilla ice cream?"

"Whoa!" Christian said. "I come to all the family get-togethers. I've only heard you say two words since you were born! Let me answer your questions. We're on another mystery hunt. No one pays us. We're staying in town right now. We don't go searching for gold and diamonds, but Sarah would love one of our mystery hunts to involve gold and diamonds. We don't get to ride horses or fly planes, but Sarah and I always fly kites. I'm not sure if we'll be on television again. When you get older, I'd consider you going with us. **My favorite dinosaur is the Ankylosaurus, and its name means '_______________ lizard'.** I only eat vanilla ice cream and vanilla cake. Any other questions?"

Henry shook his head, smiled, and ran over to ride the go karts.

Kennedy introduced two of her friends to Sarah and Christian. "These are my two best friends, Vanessa Nova and Mary Cartwright. I watch them when their parents go out." Kennedy whispered, "They don't like the word *babysit.*"

Just then, Vanessa's brother Pap walked over to the group. "Hello, my name is Papillon Nova, Pap for short. I'm Vanessa's brother. Are you friends of Kennedy and Alex?"

Christian said, "They're our cousins. It's nice to meet you and your sister and her friend Mary."

Pap waved to his sister and her friend, "Come on, Vanessa and Mary, we need to leave. I promised Mary's mom I would drive her home."

Vanessa asked, "Can't we stay longer, Pap?"

"I have to return to college tomorrow to pick up my clothes and books, so we need to leave." Pap looked at Christian and Sarah and said, "I just graduated from college and am looking for a job here in town. It was nice to meet both of you."

After Pap, Vanessa, and Mary left, Sarah said to Christian, "Vanessa and Mary must be the two students Ms. Parvana had in her class. Remember, the two girls who are so amazing at math?"

Christian said, "They're sweet girls, and I'm glad Kennedy is babysitting Vanessa and Mary. Kennedy loves to earn money to buy her jewelry and shoes." He chuckled, "She puts a little money away for college. Are you ready to walk home?"

"I am. It's been a busy week."

◆ ◆ ◆

The following Monday, the friends met at Sarah's house.

______________ said, "I was checking online for other stolen math objects, but nothing yet."

______________ said, "The suspect probably started his collection. We need to check online every few days."

Sarah said, "Yeah. It could just be the beginning of the thefts. But it could be the end because maybe he got what he wanted."

______________ said, "I read several online comments. Book lovers and butterfly lovers are upset over the loss of these objects, encouraging the person to return these items. Nothing stands out."

"I just had an idea," Christian said. "Why don't we look online for art and book thefts over the past several years? Putting aside those that just wanted to sell the items, we could create a profile of someone who collects math objects."

______________ said, "Genius!"

______________ said, "In the next few weeks, we should research past art and book thefts from around the world."

"I agree," Sarah said. "I remember reading articles on art thefts but never book thefts."

______________ said, "We need a break."

______________ said, "There's a concert tonight. We do need a break. Do you want to meet downtown?"

The friends agreed to meet downtown in the park.

◆ ◆ ◆

There were so many people at the concert. The town had decorated the gazebo with sparkling red, white, and blue lights. An orchestra was on stage, with musicians playing violins, the piano, trumpets, trombones, and guitars. The orchestra played *The*

Musical Offering, by Bach. There was a band named *PHI 21*. The music was light rock and jazz. The four friends looked at each other in amazement!

Sarah said, "I'm sure we're thinking the same thing."

"Yep. We are studying *Phi*, the basis for *The Golden Ratio*. *Phi* is the 21st letter of the Greek alphabet. The band's name is *PHI 21*." Christian asked, "Could it be a good luck sign?"

They smiled and sat on a blanket, enjoying the music. Sarah saw Kennedy, Alex, and the two little girls, Vanessa and Mary with Pap, and Kennedy's parents sitting with Ms. Parvana. They waved and smiled. The friends walked home. The mystery hunt would continue tomorrow.

THE GOLDEN RATIO

Your name: ________________________ Friend's name: ___________________________
Your favorite song: ______________________ Musical artist's name: ________________

CHAPTER THREE

IT'S ALL RELATIVE!

Sarah thought of her kite string as a lifeline tethering her to the earth while the universe was tugging at her, planning her escape. Her dreams transported her to the heavens and back, condensing time and space. She thought about the invisible force. She became fearful for a moment but was comforted as she watched Christian struggle with his kite. He always brought her back down to earth.

Christian and Sarah spent the day at the park. Christian flew his silver number 7 kite, and Sarah flew her butterfly kite. It was a beautiful day with a strong breeze - a perfect day for kite flying and a perfect day to solve a mystery.

They talked about the mystery hunt. Christian said, "There's no evidence of the sale of the butterfly and books. We need to continue researching other missing mathematical objects and follow other leads."

"I know it's frustrating, but we don't give up. Are you planning anything special this week?" Sarah asked.

"I'll keep searching online posts. I received emails from each of you. We researched past thefts over the last five years, and they prove our theory. The suspect might sell the family heirlooms."

Sarah felt a tug on her blouse. She turned around, and there stood Vanessa and Mary holding kites.

"Hello, I remember you two cuties," Sarah said. "You're Kennedy's friends."

Vanessa said, "We remember you, too!"

Sarah asked, "Are you having a fun day?"

Mary said, "Yeah. We love your kite."

"I remember, Ms. Parvana told me you both love math and butterflies. I love math and butterflies, too!" Sarah said. "Who's here with you?"

Vanessa said, "My brother, Pap. He likes math, but I like math and butterflies."

Sarah said, "Christian is my relative. He's my cousin. Pap is your relative. He's your brother."

Christian smiled and said, "Everything's relative."

Sarah rolled her eyes and smiled at Christian.

Just then, Pap ran up to Vanessa and Mary. Pap said, "Sorry, I saw them run towards you, but they can run fast."

"Hello," Christian said. "Vanessa was just telling us how much you love math, but she loves math and butterflies."

"Yes," Pap said. "Vanessa loves math and decorates everything with butterflies. I just finished college with a major in Math and Mechanical Engineering and am searching for a job."

Sarah asked, "What does a mechanical engineer do?"

Pap answered, "Well, many things. They help design spaceships, automobiles, power plants, and machine equipment. I'm focusing on research and development."

Sarah said, "Sounds complicated, but interesting."

Pap picked up Vanessa, held her close, and twirled her around in his arms. Vanessa laughed and giggled.

Christian said to Mary, "I can pick you up and twirl you around in a circle."

Mary squealed, "Yes, please!"

Christian twirled Mary around and around, and she smiled and giggled.

Both girls were dizzy, so they sat on the grass and continued to hold their kites, blowing in the strong breeze. Pap opened a blue bag and pulled out two containers of bubbles. He held their kites and handed bottles of bubbles to Vanessa and Mary. They blew blue and purple bubbles. A couple of bubbles landed on Vanessa's nose, and she giggled. Everyone smiled and enjoyed a warm summer day.

Sarah whispered, "A bright, breezy day is perfect for kite flying and blowing translucent bubbles!"

Vanessa hugged her brother Pap and said, "I love Pap."

Pap responded with a hug and said, "I love you more than the sun, the moon, and the stars!"

Pap, Vanessa, and Mary took their kites and continued to play in the park. Sarah and Christian walked home.

Sarah said to Christian, "I love making new friends."

Christian smiled.

◆ ◆ ◆

The next day, the three friends walked to Sarah's house.

Olivia answered the door. "Hello, come on into the dining room. I have donuts and juice for the four of you."

Sarah was at the table with her laptop open. "Hi! I've been searching the Internet, looking for clues."

_____________ said, "I found something this morning. An item is missing from the science museum. The object is on loan from the Einstein Museum."

_____________ said, "Well, just as we predicted, someone's on a mission to collect math objects."

Olivia said, "Even though you're young, I'm sure with your love of *all things math*, Albert Einstein is a familiar name."

"Yeah," Christian said. "One of the greatest minds and greatest mathematicians in the world. Albert Einstein developed the *Theory of Relativity*. He's best known for his mass–energy formula $E=mc^2$ or E=mc squared, the world's most famous equation!"

Sarah smiled and laughed, "Earlier today at the park, Christian said, 'everything is relative.'"

Olivia asked, "What's missing from the science museum?"

"A photograph of Albert Einstein next to his mathematical equations on a chalkboard," ___________ said. "It looked like a copy of the original, so it's probably not worth much. I'm sure the suspect took the photo because it's a math object."

"Yeah," Christian said. "Whoever's involved, I'm sure this person is collecting these objects for a personal reason and has no plans to sell the objects. Some of these artifacts aren't worth much and others are worth a lot of money."

______________ said, "Let's go to the library tomorrow and research everything we can find on Einstein and the object."

_____________ said, "I'll keep checking online to see if any other things are missing."

◆ ◆ ◆

The next day the four friends met at the library and sat at their favorite enormous oak table. The friends pulled books on Einstein and his theories.

______________ said, "There's an entire section of books on Einstein!"

Christian read, "Albert Einstein was born in Germany, in the year _____. Later in life, he wrote about two wonders that interested him as a child. At age five, he became fascinated when looking at a compass with its 'invisible forces' moving the needle. **The second wonder was when he was 12 and read a book on ____________."**

Christian smiled and continued to read, "When Einstein turned 16, a friend of his father's gave Einstein a series of popular physical science books. He imagined himself 'riding along with electricity inside a telegraph wire.' Einstein envisioned running along with a beam of light. He thought if light is a wave, it would be still."

The four whispered, always using their 'indoor' voices. They watched several small children walking in a line to a colorful room decorated with pieces of construction paper tacked to the blue walls. Each paper was designed with children's handprints. The children were excited about the puppet show. The teacher instructed them to stop, and they stood still, waiting to move forward.

___________ said, "They're so adorable. It reminds me of Einstein, a line or a light beam moving along, and then looking still."

The friends smiled. They took a break and walked over to the room to watch the small children enjoying the puppet show. The dolls with strings were colorful, and they were teaching the children about famous mathematicians. One puppet resembled Einstein, and the friends looked at each other in amazement! One doll was named Mary Cartwright! They watched the children laugh and giggle and learn about brilliant mathematicians.

Sarah asked, "There's a famous mathematician named Mary Cartwright? I'm sure that's the name of Vanessa Nova's friend, whom Kennedy babysits. Ms. Parvana is Mary's teacher. She told us Mary and Vanessa love math. I need to research Mary

Sarah whispered to Christian, "Einstein was interested in an 'invisible force.' I'm sure it's not the same invisible force we're dealing with on these mystery hunts." Cartwright in one of these books."

_______________ said, "Well, the puppet show is on famous mathematicians, and one puppet is Einstein, and we're researching Einstein. Crazy!"

Christian and Sarah walked over to the bookshelves.

Christian said, "I'm sure!" He paused and said, "I found a book on famous female mathematicians."

Sarah opened the book and read, "Mary Cartwright is one of the first mathematicians who studied ________________ ________________."

Sarah and Christian looked at each other and didn't speak for a few minutes.

"This can't be a coincidence!" Christian asked, "Is it a sign? We were studying *The Butterfly Effect* and Professor Lorenz when Ms. Parvana described how much Vanessa and Mary love math."

The other two friends overheard Christian and joined Christian and Sarah back at the table.

_______________ said, "Speaking of signs, I planned on telling you this later today. The name Vanessa and the last name Nova, both mean butterflies."

The friends were silent for a few minutes.

_______________ asked, "I wonder what Pap's name means?"

"I don't know. But, what's the famous quote in the Shakespeare play, *Romeo and Juliet*? 'What's in a name?'" _____________ asked. "We should ask this question every day. All of this is giving me chills."

Sarah added, "I need some time to go over everything we've learned and how it relates to what's happening. Why don't we come back to the library tomorrow at 1:00 and continue researching Einstein?"

Christian said, "We each need a day to process this information. There are no coincidences."

◆ ◆ ◆

The friends walked home; none of them said a word for several minutes. When they decided to speak, they talked about the names and their meanings, and if this was a sign.

Sarah arrived home. Olivia was working on her laptop. Sarah said, "I have to talk to you. Something strange happened."

Olivia closed her laptop. "I hope you're safe. What's strange?"

Sarah said, "We found several coincidences. But we don't believe in coincidences. We haven't shared our special dreams with our friends. I haven't told Christian yet, but I dreamt of butterflies a few nights ago."

Olivia asked, "What did you dream?"

Sarah said, "*I was in a little girl's bedroom. I couldn't see her face, but she was twirling around and wearing a butterfly dress, butterfly sneakers, and a butterfly hair clip holding up her brown curls. There were seven butterflies, but they could only fly so high. The butterfly on the hair clip resembled the missing framed butterfly. I asked her name, and she whispered, 'Butterfly.' There were three books on a shelf in her room, one book with spirals and butterflies on the cover, one book on relatives, and one book with the title '*Symmetry.*' I grabbed the book with the spirals and butterflies, but when I looked down, I saw the word '*Relative.*'* My dream must be about the missing butterfly."

Olivia said, "A few years ago, I justified your dreams as just images of things you saw during the day. But Christian dreams too, and these dreams must be signs. Kennedy and Alex have these special dreams. Maybe the 'Relative' book is about our family. You, Christian, and Kennedy and Alex are related and have unique dreams."

Sarah said, "Or it could be about Einstein. One missing object is a photograph of Einstein. He discovered *The Theory of Relativity.* When we started this mystery hunt, we met with the owner of the framed butterfly, Ms. Parvana. She teaches elementary school-aged children with autism and mentioned two of the children love math. We met both little girls at Alex's Congratulations Party.

"Kennedy babysits Vanessa Nova and Mary Cartwright. To make a long story short, today, we realized Vanessa's first and last name, each means *butterfly*. Mary Cartwright was the name of a famous female mathematician who studied *The Butterfly Effect*."

The sisters were quiet for a few minutes.

Olivia asked, "Are you saying either Vanessa or Mary could be the little girl in your dream?"

Sarah answered, "One of them could be. It might be someone I haven't met. The little girl could be a symbol of the missing butterfly. I'm confused."

Olivia suggested, "You always write your dreams in your notebook. Share your dream with Christian, and ask him if he had a similar one."

"Okay. This mystery hunt is getting stranger by the day. On a positive note, we're learning facts on famous mathematicians and their theories," Sarah said with a smile.

Olivia said, "I always wanted to research the lives of brilliant mathematicians, scientists, and artists, to discover the people who motivated them to choose their paths."

◆ ◆ ◆

The next day, Christian knocked on Sarah's door and asked, "Ready to go?"

Sarah said, "Yeah, but I have to talk to you. I had one of my dreams. But I'll tell you about it while we walk."

As the two cousins walked to the library to meet _______________ and _______________, Sarah told Christian her dream.

"I don't know what to make of this," Christian said. "I haven't had any of my dreams lately."

Sarah said, "We should keep our dreams to ourselves, at least for now."

"I agree."

As Christian and Sarah walked into the library, they saw _____________ and _______________ sitting at their favorite table with books open.

"We continued researching Einstein, the photograph, and his theories," ____________said. "We're trying to connect the names. I couldn't think of anything else last night."

_______________ said, "Yeah, me too. These images floated around in my head and in my dreams."

Sarah and Christian looked at each other and smiled.

Christian said, "For now, why don't we accept the names as sort of interesting connections? Even though these connections are probably more than just chance."

"I agree," Sarah said. "Let's keep researching Einstein and his theories."

________________ read from one book on Einstein. "**Einstein took a job in a Patent Office after leaving Germany for ___________________.** He continued to dream of racing along next to a beam of light. **While he was at school, he studied James Clerk Maxwell's theory on _____________________________.** Einstein continued his search for answers. He formulated that the speed of light remains the same no matter how fast a person or object moves. **This understanding led Einstein**

to complete his General Theory of _______________ in November 1915. Einstein and the world considered this theory to be his greatest discovery."

________________ said, "I watched a program on Einstein once. He was so smart! $E = mc^2$ is Einstein's most famous equation. E, or energy, represents the total energy of a system; m, or mass, is related to energy; and c^2 is the speed of light squared. The narrator said that this is the correct formula to make mass and energy equal."

"The photograph is important!" Sarah said. "This is a picture of Einstein near a blackboard where he wrote an equation."

"It's probably only a copy. Anyone could pull the photo off the Internet," ____________________ said. "This photo must be special in some other way."

Christian said, "Let's research the online article on the theft."

_________________ opened the laptop and typed. "Give me a few minutes to reread the article online. Sorry, but my laptop's out of power. I forgot to bring my power cord."

Christian opened his laptop. He said, "My computer's out of power too, but I charged it before I left home."

Sarah and Christian looked at each other. They knew they were thinking the same thing ... the invisible force.

_______________ said, "That's weird! We can use one of the desktop computers in the library."

The friends went from computer to computer. Someone was sitting at each computer. They came to the last computer near the back door. There was a sign on the computer, "Out of order. See the librarian at the front desk."

The friends went to the front desk and asked if there was a computer to use. The librarian pointed to one in a separate room. She said, "We started having trouble this morning with a couple of computers, but I just used that one. It works."

____________ turned on the computer and started typing. "Okay, this one works. Einstein signed a copy of the photo. Let me read the inscription on the back. You won't believe the inscription!"

Sarah said, "Read it! Read it!"

"To my dear little butterfly, Lieser. Always wish, Lieserl, always LOVE, hope, and dream."

"I read somewhere he only had one daughter, but she passed away when she was a baby," Christian said. "He also had two sons. This must be the reason the photo's so valuable."

Sarah said, "But he called her butterfly. This can't be a coincidence!"

The friends nodded.

"It's getting late in the day. We should go home," ________________ said. "I'm hungry, so it must be dinner time."

Christian and Sarah walked home together. Christian asked Sarah, "What if the little girl you were dreaming about is Einstein's daughter, Lieser or Lieserl? And why did he spell her name differently on the back of the photo?"

Sarah said, "I'm not sure. But what about your laptops and a few computers in the library, not working? Very weird!"

Christian said, "The computers not working. Mr. Bigollo's car not starting with the tree branch falling and almost hitting his car. He could have been really hurt. All of these things made me think about Tata's warning. You're right … very weird! But we're not going to let someone or something stop us. We also have to play this out and find the person and the missing items. Then we can uncover what's going on with all these names. I want to know how these names are related to the missing objects."

Sarah asked, "What if we find the person and the stolen things, but we don't have an answer to my dream, the names, and the connections? I also don't want anyone to get hurt by this invisible force if there is such a thing. It scares me a little."

Both cousins were quiet as they walked home. Neither of them had an answer.

◆ ◆ ◆

Christian called his Uncle Raymond, who asked, "How's your mystery hunt going?"

Christian said, "It's becoming a greater mystery every day. Another object is missing. We read about it online."

"Yes," Uncle Raymond said, "the photo of Einstein. We're looking into it. The Science Museum Director, Dr. Theis, is out of town."

"We're experiencing weird things that seem like coincidences. Sarah had one of her dreams. I haven't had one yet. We depend on these dreams to give us clues on our mystery hunts and guide us to the answers," Christian said. "Could I meet with you?"

Uncle Raymond asked, "Are any of you in danger?"

"Yes, and no," Christian answered. "It's not about feeling unsafe. Weird things have been happening. There are strange things we can't explain about people's names. It's as if these are signs from someone trying to give us clues. Also, Tata warned us about an invisible force trying to stop us from helping, but I can explain that to you. Can we meet at Sarah's house tomorrow, just in case she has more info?"

Uncle Raymond said, "Sounds good. Text me with the time, and I'll be there."

Christian called Sarah, and they agreed to meet with Uncle Raymond the next morning.

Sarah said to Christian, "I need to tell you something I found when researching Einstein, but it can wait until tomorrow."

◆ ◆ ◆

The next morning Uncle Raymond met with Christian and Sarah.

Sarah said, "Before we start, Christian, because of the 'no-coincidences' with the names, I looked up Einstein's daughter's name, Lieser. Lieser is a German name and means 'love' and 'of books,' or 'the love of books.' Two of the missing objects are books. We've been reading books at the library, and there were books in my dream. Lieserl is a German name meaning 'God is my oath'. I think someone's trying to send us a message."

Christian and Sarah wrote everything they learned on a sheet of paper for Uncle Raymond, including the names, and Sarah's dream.

Christian said, "Tata had our special gift. She warned us more than once about an invisible force that would try to stop us from solving these mysteries."

Uncle Raymond responded, "This is well above my pay grade or my understanding. I've never been a believer in coincidences or invisible forces. But my years as a detective have taught me many things. What seemed like coincidences were actually valuable clues in solving investigations. If there's a message, it's a reminder of our connections to each other and how important it is to work together. Sarah, the book in your dream, had the word *Relative* on the cover.

"The word relative may be connected to your research on Einstein's Theory of Relativity. **Symmetry is a ___________ theory, and the first object taken was a framed butterfly.** You learned about *Chaos Theory* or *The Butterfly Effect*. Your family has special dreams, and they're fascinating. Both of your dreams helped in our search for and recovery of Professor Septenary. Please remember safety first. And as far as Tata's warning ... she wouldn't warn you about something she doesn't believe in."

Christian said, "Thanks, Uncle Raymond. We'll text you about anything we discover."

Sarah said, "Yeah, thank you. Our dreams are confusing in the beginning. But, as we continue deciphering clues, our dreams point to people and places, and sometimes the answers of how or why. We're all pretty curious, and that's a good thing. We won't stop looking for answers. Next week we're meeting again, and we'll keep researching missing math objects."

Uncle Raymond said, "I'm not allowed to share confidential information, but I'll share with you what I can. We worked well together in the search for Professor S."

"We're meeting our friends a few times a week, and we're researching math objects online," Christian said. "The more objects the suspect takes, the more info we have."

Uncle Raymond said, "I can't tell you much else, but this person's not leaving strong clues behind, so for now, we haven't identified the suspect. Once again, stay safe. And whether or not you believe there's an invisible force trying to stop you, if any of you feel in danger, stop the investigating. We'll take it from there."

ALBERT EINSTEIN

CHAPTER FOUR

VIOLINS, STRINGS, AND EVERYTHING

Christian woke from a deep sleep. He called Sarah, "Now, I had a dream of butterflies, stars, and little girls! Just like you!"

"In my dream, *a small bedroom was filled with butterflies, books, and musical instruments. Butterflies were all around me but I wasn't flying, just floating mid-air, and I noticed three small girls. One girl played an antique violin, plucking the strings and giggling. Another girl was holding a harmonica. She started to play the harmonica, putting it up to her lips, but then the room's ceiling fell away. The girls looked up in amazement and saw the universe, numbers, and stars floating in a deep night sky. The third girl sat at a small table with four chairs until seven butterflies floated down and landed on her dress and sneakers. One butterfly landed in her hair. Then, someone scooped the numbers and stars out of the sky with a butterfly net and then floated into the room. The little girl with the butterfly in her hair opened her hands, and this person dropped the numbers and stars into her cupped hands, and she giggled. Then a 'shadow hand' reached out to grab books from the shelf, and the numbers and stars. I couldn't see whose hand it was, but it frightened the little girls. Then I saw the same 'shadow hand' closing a door with bars and slamming it hard."*

Sarah said, "Wow!" and then paused for a minute and said, "I'm planning on babysitting our three cousins next week. The little girls in your dream could be Marisa, Anita, and Ameliah."

"I'm not sure," Christian said. "I didn't see their faces, but the one girl with the butterfly in her hair could be the girl in your dream."

"Well, your dream has three little girls, a couple of mysterious characters, and musical instruments," Sarah said. "This will help us on our mystery hunt. Did you see the face of the mysterious character scooping the numbers and stars out of the sky? Could this character be you? You're trying to solve the mystery of the missing math objects. Do you think the mysterious hand was the same person trying to scoop the stars and numbers out of the sky? Could the 'shadow hand' be the invisible force?"

"I didn't recognize anyone. All I could think about was the invisible force. But I wasn't scared, the girls were afraid, especially the little girl with the butterfly in her hair.

"So, the next object should have something to do with math," Christian said. "**Not unless these suspects are changing their ___________ Operandi or M.O., changing their habits from taking just math objects.** My guess is the next missing object will be a violin or a harmonica."

Sarah asked, "Was this person in your dream male or female?"

"I could only see a silhouette of a person," Christian answered. "For all I know, it might be an angel! Remember, angels don't have wings."

Sarah replied, "Yeah, only Seraphim and Cherubim have wings. They're not angels. Angels are messengers."

Christian remarked, "Nice, Sarah."

Sarah smiled and said, "I hope that between our two dreams, these images and clues will help us find the objects and return them to their owners."

"I hope so too," Christian said. "We're going to meet at the library. See you soon."

◆ ◆ ◆

The four friends met at the library and sat at their favorite table.

____________ said, "I've been checking online every day. I read about another missing math object!"

Sarah asked, "Was it a violin or harmonica?"

Christian smiled at Sarah, and the other two friends looked puzzled.

____________ asked, "What's up with a violin or harmonica?"

Sarah smiled at Christian and said, "Just a guess."

"Yes, and no," ____________ said. "A painting from the planetarium is missing."

____________ asked, "What?"

"The article online talked about an artist creating the painting, *Violins, Strings, and Everything*," ____________ said. "The artist painted a violin and butterflies in an hourglass."

Sarah asked, "Why was the painting in the planetarium?"

"The article said the Planetarium Director has degrees in astronomy and physics. The painting is from her collection," ____________ said. "It hung in the lobby before her office. Anyone visiting the planetarium could have taken the painting."

____________ asked, "Your guess was right, but how could you be so sure the next object would be a violin or harmonica?"

Sarah said, "It was just a guess, and it's a long story."

____________ said, "I'm sure your Uncle Raymond's on the case."

Sarah asked, "When was it taken?"

____________ said, "Just two days ago."

____________ asked, "What's *String Theory, again*?"

Christian answered, "*String Theory, Superstring Theory*, or *The Theory of Everything* is a math concept."

____________ asked, "*The Theory of Everything*? Is there an idea to explain everything?"

Sarah answered, "We'll research these theories. I planned on helping my mom today, so Christian will text you about meeting again."

"Sounds good," Christian said. "I'll give my uncle a call tonight. I'm sure he knows a painting is missing."

The friends left the library to walk home. Sarah and Christian needed time to talk alone. As they walked on, Sarah and Christian started talking about Christian's dream.

◆ ◆ ◆

Sarah said, "The painting has a violin with strings, butterflies, and an hourglass. I guess the hourglass represents time. Time is a mathematical idea. You dreamt of a violin in your dream, strings on the violin, *String Theory*, and butterflies and strings of musical notes. The person dropping the stars and numbers into the little girl's hands could be the owner of the painting or the planetarium director."

Christian said, "I didn't hear about this missing painting before dreaming. Our dreams are giving us clues about the missing objects, and the suspects. I'll call Uncle Raymond, but for now, I don't want to tell anyone else."

"I agree. It would be best if Uncle Raymond called the planetarium director. He could schedule an appointment with her and explain about the other missing objects."

Christian said, "Yeah, I'll text you tonight. See you tomorrow."

◆◆◆

Christian called his Uncle Raymond after dinner.

Uncle Raymond said, "Hello, Christian. How are you?"

"Great, thanks. One of our friends read online about another math object missing, a painting."

"Yes," Uncle Raymond said. "The missing painting's a creation of the math concept *String Theory*. I attended high school with the planetarium director, Aurora Draco. She's extremely knowledgeable and very gracious. I met with her yesterday and she'll wait for your call. She loves teaching students," Uncle Raymond said. "Dr. Draco's excited you offered to help and to teach the four of you about *String Theory*. It was her focus in college for her Physics degree."

"Thank you," Christian said. "I had another one of my dreams last night. I'll share a little bit with you. *There were three little girls, one was playing the violin, and one held a harmonica in her hands. Another person was in the room, but I wasn't able to tell if the person was a woman or a man. This person scooped stars and numbers out of the sky and dropped them in the first little girl's hands. Then a 'shadow hand' reached out to grab books from the shelf, and the numbers and stars. I couldn't see whose hand it was, but it frightened the little girls. Then I saw the same 'shadow hand' closing a door with bars and slamming it hard."*

"Amazing!" Uncle Raymond said. "I know that several members of your family, over many generations, were gifted with the ability to dream prophetic dreams. You and Sarah will figure it out. Did you hear about the missing painting before your dream?"

"No, ____________ read the article to me at the library."

There was silence over the phone. Uncle Raymond remarked, "Thanks for sharing your dreams with me. I'm fascinated by what they could mean. Please give Director Draco a call tonight and schedule your meeting. The four of you should go. It'll be educational and informative."

"Thanks again, Uncle Raymond."

◆ ◆ ◆

The next morning, Christian called The Strasenburgh Planetarium and spoke with Director Draco.

"I spoke with your Uncle Raymond yesterday. He's an old friend and schoolmate. I'm grateful you'll help in the search for my painting," Dr. Draco said. "I'm happy to explain *Superstring Theory* to you and your friends. I wrote my dissertation on this theory."

"Thank you," Christian said. "We're looking forward to meeting with you. What's a dissertation?"

Dr. Draco answered, "A student completes a Thesis for a Master's Degree and a Dissertation for a Doctorate Degree. A dissertation is a very long research paper. The student contributes knowledge and theories to their field of study. In my case, it was for my Physics degree."

"So," Christian said, "may we call you Dr. Draco?"

"Yes," Director Draco said. "I'm not a medical doctor. I'm a doctor of physics."

"Thanks for explaining everything to me," Christian said. "The four of us love to learn, and we're looking forward to helping in the search for your painting. I'm excited to learn about *Superstring Theory*."

◆ ◆ ◆

Early in the evening, Christian emailed Sarah and their two friends and asked them to meet him at the planetarium at 11:00.

The next day the friends walked into the planetarium. Dr. Draco stood at the door and waved to them.

"Good morning. It's so nice to meet you. I almost didn't make it this morning. The strangest thing happened. My front door wouldn't unlock, and then I couldn't find my keys," she chuckled. "I found them in the refrigerator, and I can't imagine how they got there. I came home late last night and didn't open my refrigerator. I had to go out the back door to my car. But I have a locksmith coming to my home to fix the door later today."

Sarah looked at Christian and said, "That's strange."

Christian nodded and said, "Thank you for meeting with us, Dr. Draco. This is my cousin Sarah, and our two best friends, ________________ and _______________."

"Welcome. Please follow me to my office."

The friends followed Dr. Draco to a long hallway filled with paintings on the walls. There was one space for the missing *String Theory* painting.

As the friends walked into the sizeable office, there was a radiant stained-glass window behind Dr. Draco's desk. On the four walls were butterfly paintings, a night sky filled with stars and planets, and many artist renderings of *String Theory*. There were enormous bookshelves on the walls, filled with many books. Sarah recognized the music playing, a violin concerto. It was one of her mom's favorite classical compositions.

Dr. Draco invited the friends to take a seat, "Please make yourselves comfortable."

Sarah asked, "Dr. Draco, do you have a photograph of the missing painting?"

"Yes," Dr. Draco said. "I made copies for you." Dr. Draco handed the copies to the friends.

Sarah said, "It's lovely!"

__________________ asked, "Do you have any idea who took your painting?"

"No, I don't," Dr. Draco said. "The day it went missing, we had hundreds of children, parents, and teachers, along with other visitors touring the planetarium. Many children from several schools had created art displaying their thoughts and dreams of the universe. As you can see from the empty spot on the wall in the hallway, the painting was up high so that no young child could reach it and pull it down."

Sarah asked, "May I ask you a question? It's a little off-topic."

Dr. Draco said, "Yes, of course."

"You have an amazing collection of butterfly paintings," Sarah said. "Our mystery hunt began with a butterfly. Why do you collect butterfly paintings?"

Dr. Draco smiled and replied, "I have always loved butterflies. I became interested in math during my first year at college. I wasn't sure what to choose as my major at the university. But then I listened to one of my professors describe *The Butterfly Effect.* Have you heard of this theory?"

The friends smiled and nodded.

Christian said, "We researched this theory when we met the owner of the missing framed butterfly. Professor Lorenz once owned it."

"Wonderful!" Dr. Draco said. "Learning expands our minds and enriches our souls. Now, let me have the privilege to explain *Superstring Theory.*"

Dr. Draco continued, "We call *Superstring Theory* 'The ________ __ _________.' Scientists around the world continue to study this theory. They hope to reconcile or merge the differences between *Quantum Theory* and *The Theory of Relativity*, to explain the nature of known forces and matter. *Quantum Theory* is a theory of matter and energy based on the concept of quanta or a quantity of energy. ***The Theory of Relativity* is a theory proposed by ____________ ______________."**

Sarah said, "We've been lucky to learn facts on *The Theory of Relativity* on our mystery hunt. We appreciate everyone taking the time to teach us along the way."

"The field of Mathematics touches each of our lives," Dr. Draco said. "Even as a child, I gazed at the night sky and dreamt of stars, numbers, and butterflies floating in the heavens."

Sarah and Christian looked at each other and smiled.

Christian said, "One thing we're learning is that everything's connected. Could this be the answer to *The Theory of Everything*? Professor Septenary's favorite quote is, 'When we try to pick out anything by itself, we find it hitched to everything else in the universe.' Sarah and I believe the answer to the mystery of the universe is Intelligent Design."

Sarah nodded in agreement.

Dr. Draco smiled, "In my opinion, the complexity of nature, and the intimate design of our universe must be by the Hands of God. I appreciate students working, dreaming, and most importantly, considering our very existence is by design. Our world is a better place when we embrace our connections and our differences."

Christian said, "Thank you, Dr. Draco."

Dr. Draco continued, "***The Theory of Everything* or its abbreviations, __ __ __, is the final theory, or master theory of a single framework of physics that explains and links all physical facets of our universe."**

________________ asked, "Did someone give you the painting or did you buy it?"

Dr. Draco said, "A dear friend of mine gave it to me for my birthday. She plays the piano, violin, saxophone, guitar, harp, flute, and even harmonica. Her background's in physics, and she loves butterflies, too!"

Christian and Sarah looked at each other. Sarah asked, "A harmonica and a violin?"

"Yes," Dr. Draco said. "Are you aware the harmonica has another name? **We call it a ___________, not to be confused with a ___________ with strings.** As children, my friend and I chased butterflies with our butterfly nets. She's so funny. We spent many days jumping very high to catch stars and butterflies with our nets."

Sarah and Christian looked at each other again and smiled.

Sarah whispered to Christian, "The painting is titled, *Violins, Strings, and Everything*. Dr. Draco's speaking about the *Theory of Everything*! Could your dream be about Dr. Draco and her friend?"

Christian smiled and nodded.

_____________ said, "Thank you for the information. Are you missing any other objects?"

"No, just the painting. There are other objects much more valuable than the painting. But this painting is precious to me," Dr. Draco said. "It was a gift from a dear friend, and represents *String Theory* and its connection to music, our universe, the *Butterfly Effect*, and time."

Christian said, "We'll email you if we find information on your missing painting. Did my uncle explain there are other objects missing? A framed butterfly, two books, and a signed photograph."

Dr. Draco said, "He explained the missing objects have math connections."

"Thank you again," Sarah said. "We appreciate everything you've shared with us. We've learned so much!"

"You're welcome. Please visit us again. I'll give you a tour of the new wing of the planetarium."

◆ ◆ ◆

The friends walked home.

They took a few days off from the mystery hunt to spend time with family.

Olivia and Sarah drove to the movie theater. Their local movie theater, *The Brass Ring*, showcased theme films for $2 a person during weekday afternoons. The first two movies were short films, *Eustice Solves a Problem*, and *Solid Geometry*, with the main attraction *Stand and Deliver*.

As they began watching the movies, Sarah asked Olivia, "Are these three movies on math?"

Olivia said, "I read the titles and thought the first two short movies could be on math, but I didn't know the main film was about a math teacher using creative methods to teach math to gang members."

Olivia and Sarah had popcorn and bottled water at the theater. After the movies, they planned a quick lunch before driving home.

As the two sisters enjoyed their lunch, Olivia asked, "How goes the mystery hunt?"

"We're not getting very far," Sarah answered. "Someone is continuing to take objects. Christian had a waking dream."

Olivia asked, "What was his dream?"

Sarah said, "He dreamt t*here were three little girls, another person he didn't recognize, and a violin and a harmonica. This mysterious person caught numbers and stars with a butterfly net. He also saw a 'shadow hand' try to grab the books, numbers, and the stars, and then this hand slammed a door with bars.* My dream was of one little girl."

Olivia said, "That's weird and a little scary. Did both of your dreams have butterflies?"

"Yep. Christian only dreamt of an outline of a person who scooped numbers and stars from the sky with a butterfly net and dropped them in one of the little girl's hands. He didn't know if it was a man or a woman. But he's sure the person scooping the numbers and stars with the butterfly net is a different person than the figure with the 'shadow hand.'"

Olivia asked, "What does it mean?"

"Christian's dream may have pointed us to the next missing object, a painting from the planetarium. The painting is of a colorful violin, butterflies, strings, and an hourglass. The Planetarium Director, Dr. Draco, owned the painting. A good friend gave her the painting titled *Violins, Strings, and Everything*. Dr. Draco told us a nice story about it. She and her best friend played in the park. They tried to catch butterflies with a butterfly net and scoop up stars and numbers in the sky, just like the little girl in Christian's dream."

"Your dreams give you clues to solve the mystery. They always amaze me! Kennedy and Alex have this special gift too, and sometimes, not all the time, I wish I had this gift as well," Olivia said. "I studied *The Theory of Everything* in college. Professor Septenary taught this subject. **He discussed the particles of our universe, including __________, fermions, leptons, and quarks.** He explained these particles are *the letters making up the words* to explain matter and its space.

"*Superstring Theory* is a math theory designed of loops. **The loops are compared to an infinite rubber band or vibrating violin ___________.**"

Sarah said, "Strings, violins, numbers, stars, and butterflies. Where does the harmonica fit in?"

Olivia said, "Well, I'm not sure, but all music creates vibrations. I'll ask Professor S; he taught *String Theory*."

"Thanks," Sarah said. "We have more questions than answers on these mystery hunts."

Olivia asked, "Ready to go home?"

Sarah said, "I wonder if Professor S is still at the college."

"I'm not sure," Olivia said. "He retired, but now he's teaching a couple of classes. I'll drive there and ask if he's available to meet with us for a few minutes."

◆ ◆ ◆

As soon as the two arrived on the campus of Heptad University, they spotted Professor S sitting outside on a bench reading the book, *Symmetry*. The two sisters waved to the Professor.

Professor S waved back. "Well, hello. What a pleasant surprise! How are you?"

Olivia answered first, "Wonderful! It's so good to see you. When did you get back from your trip?"

Professor S answered, "A week ago. I stayed with my son for two months. I came out of retirement to teach two classes. It keeps my brain sharp, and at my age, it's important to instruct, learn, reflect, and dream."

Olivia asked, "What are you reading?"

Professor S said, "A new book on *symmetry*."

Both sisters looked at each other and smiled.

Sarah said, "We're so glad you spent time with your son. I hope you don't mind if we ask you a few questions."

Professor S said, "I'm happy to help. Sarah, are you on another mystery hunt?"

"Yes," Sarah said. "You'll love this mystery hunt. It's all things mathematical."

"I love all things math, particularly the number seven," Prof. S said with a wink and a smile.

Olivia and Sarah looked at each other and smiled, and Sarah added, "...and continents."

Sarah said, "Several math objects are missing, and we're trying to help the owners locate them."

"I read an article in the newspaper about the missing objects. It sounds as if someone has a fascination with mathematical objects," Prof. S. said. "How can I help?"

Sarah said, "We're studying *String Theory* and any connection with musical instruments. I understand how violin strings might be connected to this theory. Is the harmonica connected to any math theory? Christian dreamt of a violin and a harmonica."

Professor S said, "Musical instruments, even the human voice, create vibrations. Why don't you meet me at my office tomorrow? My class begins in a few minutes, and I'll try to answer your questions."

Sarah said, "We'll bring Christian along."

"Great! I enjoy spending time with the three of you. We will talk mathematics," Professor S said. "See you tomorrow at 11:00."

VIOLINS, STRINGS, AND EVERYTHING

CHAPTER FIVE

HARMONY

Could Professor Septenary help the four friends find the clues to unlock the next mystery?

Christian knocked on Professor S's door at Heptad University.

Professor S opened the door and said, "Welcome, please come in. It's so nice to see you again. How are your friends?"

Christian said, "They're fine and helping us out with this mystery hunt again. We're the four musketeers, I guess."

Prof. S smiled. "Please sit and make yourselves comfortable. There were four musketeers, Athos, Porthos, and Aramis, the three mentioned in the novel by Alexandre Dumas. But D'Artagnan was also an important member of the group."

Christian grinned and said, "You're always teaching us, Professor S. Do any of their names mean the number seven?"

Prof. S said with a smile, "Not to my knowledge." He turned to Olivia and said, "Olivia, you're finishing your second year at the university. Are you looking forward to your classes?"

"Yes," Olivia said. "Thanks for asking. We'll miss you teaching every day, but it's great to see you around campus."

"Thank you. I'm enjoying teaching two classes. The students are great. This campus is my second, or third home," Prof. S. said. "I have three homes, one in Iowa, my home here in town, and Heptad University."

Sarah said, "Professor, after you traveled the world, I'm sure the world is your *home*. Thanks for spending time with us. We need some help in connecting our dreams with the missing objects. We've figured out the connections between the violin and the butterflies. We still need to figure out the symbolism of the two books. Christian saw a harmonica in his dream. How's a harmonica connected to math?"

Prof. S. said, "I'm reminded of a famous quote by Pythagoras, a renowned philosopher, and ____________________. 'There is geometry in the humming of the strings, there is music in the spacing of the spheres.' It may very well be the harmonica is symbolic of the mathematical theory, *Harmonic Analysis*. I pulled an article. I'll read you a quote from James Arthur's paper, 'In music *Harmonic Analysis* is the study of chords, and of how these chords are used in combination to create musical effects. This theory in mathematics takes on a somewhat different meaning. It too has roots in music, or at least in the mathematical analysis of sound.' Christian, if your dream is correct, it could connect the missing object to *Harmonic Analysis*."

Christian asked, "Are mathematicians studying this now?"

"Yes," Prof. S. added, "Mathematicians study *Harmonic Analysis* for many reasons. **The famous ____________________, Aristotle and Pluto, connected mathematics and music to astronomy, geometry, and even language.** I consider *Harmonic Analysis* as mathematical music. Composers create compositions. The musical notes and chords create a uniform piece of music. *Harmonic Analysis* is a way to understand something complex by making it simpler. We discover symmetry and harmony in understanding mathematics through music."

Christian said, "Thank you so much! This makes the theory easier to understand. We appreciate your help."

Professor S said, "You're welcome. I'm happy to help you anytime. I'm so thrilled you're continuing to learn."

"We're glad you're home," Sarah said. "You taught us how important the number seven is in religions and cultures around the world. You were our first successful mystery hunt!"

The three cousins left to return home.

◆ ◆ ◆

When they arrived home, Olivia asked Sarah and Christian, "Will the next missing object be a harmonica? What's your best guess?"

Christian said, "I'm guessing it'll be a math object connected with *Harmonic Analysis*."

Olivia asked, "Why?"

Christian said, "Only because I dreamed of a harmonica, and the missing objects have a math connection."

Just then, ____________ stopped by Sarah's home.

____________ said, "Another object's missing. Einstein wrote a theory about *Harmonic Analysis* on the back of a Bach symphony. I don't know anything about *Harmonic Analysis*."

Olivia, Sarah, and Christian looked at each other. Olivia said, "I guess you were right, Christian."

Christian said, "The three of us met with Prof. S, and he explained the theory. He connected the next missing object to *Harmonic Analysis*."

"I don't understand," ____________ said. "Why did he think the next missing object would be about *Harmonic Analysis*?"

Sarah said, "It's a long story. Where was the missing object?"

______________ said, "It was on display at our high school in the music department. The school has several musical objects on exhibit, but this is the only thing that's gone missing."

Christian said, "We'll need to talk to someone at the school."

"Miss Aria Bell's the new music teacher," __________ said. "It was her idea for the school exhibit."

"Thanks," Christian said. "Sarah, will you call Miss Bell to see if we can meet with her?"

Sarah said, "Sure."

_______________ asked, "How's Prof. S? We learned so much from him."

"He's fantastic!" Sarah said. "He asked about both of you."

◆ ◆ ◆

The next morning, Sarah called the school and left a message for Miss Bell. It was summer, so Sarah was sure it might take a few days for her to return the call.

Sarah and Christian took their three young cousins to fly kites. It was a warm, sunny day with a strong breeze.

Christian and his dad built unique, one-of-a-kind kites for Marisa, Anita, and Ameliah. Marisa's kite was the shape and colors of a ladybug, Anita's was a princess kite, and Ameliah had her favorite puppy kite. Sarah brought her purple, pink, and golden yellow butterfly kite, and Christian flew his silver number 7 kite.

The sky was blue with only a few wispy, white clouds, and the five cousins agreed it was a perfect day for kite-flying. Families were enjoying picnics and flying kites and balloons. Christian and Sarah brought sandwiches, chips, and fruit juice in a large picnic basket. After flying kites, the five cousins sat on a giant butterfly and ladybug blanket and enjoyed a delicious lunch. Christian and Sarah told some funny jokes. The three girls always loved jokes, flying kites, eating ice cream, and giggling.

Christian asked, "How do oceans say hello to each other?" The three girls looked at each other.

Marisa said, "Give us a clue."

Christian looked at Sarah, and she put her hand up in the air and moved it back and forth.

Anita shouted, "They wave!"

The three girls clapped. Anita pleaded, "Ask me another joke."

Sarah asked, "Why do bees have sticky hair?"

Ameliah asked, "Do bees have hair?"

Sarah nodded and answered, "Yes, bees have hair."

Marisa asked, "Is it because honey is sticky?"

"Close," Christian said. "It's because of their honeycombs."

The three girls laughed so hard they leaned back and rolled on the ground.

Ameliah pleaded, "One more, please?"

Christian asked, "Why did the math book look so sad?"

Marisa, Anita, and Ameliah looked at each other and said in unison, "We give up!"

Sarah answered, "Because it had so many problems! Get it; math books have math problems?"

Marisa said, "I get it, but maybe Anita and Ameliah don't get it."

Sarah said, "Okay, one more."

Christian asked, "What has hands but can't clap?"

The three girls looked at each other.

Marisa asked, "A clock?"

Sarah answered, "Yeah! A clock!"

"It's not funny," Anita said. "One more, please!"

"Okay," Christian said, "another bee joke. What haircuts do bees get?"

The three cousins looked at each other and laughed.

Marisa said, "I know, I know. My cousin had one. A buzzzzcut, because bees buzz!"

"Right!" Sarah said. "Good job."

Christian said, "You three look hungry. Ready for some ice cream?"

The girls clapped and began running over to the ice cream truck. Christian and Sarah jumped up and ran after them. Sarah ran back for the picnic basket and blanket with the blue bag holding the folded kites, as Christian caught up to the three girls.

Sarah said, "You can choose your flavor and choose a bowl or cone."

Marisa answered, "We want ice cream in bowls because we want lots of toppings."

"Okay! Order your favorite ice cream flavor and toppings," Sarah said. "You can each practice ordering. Always remember to say Please, and Thank You."

Ameliah ordered first. "Please, and thank you, I want vanilla ice cream with rainbow sprinkles, hot fudge, whipped cream, and a cherry on top, please and thank you."

Christian and Sarah smiled. Christian said, "You can say please first, and thank you after the nice girl hands you your ice cream."

Anita went next. "Please, may I have the same as my cousin? Thank you."

Marisa ordered after Anita. "May I please have vanilla and chocolate swirl ice cream with chocolate sprinkles, whipped cream, and a cherry on top? Thank you."

"Very good," Sarah said. "May I please have the same as Ameliah, vanilla ice cream with rainbow sprinkles, hot fudge, whipped cream, and a cherry on top? Thank you very much."

Christian half-smiled and asked, "May I please have a large cone of vanilla?"

Sarah smiled and said, "Shocking! Just-plain-vanilla."

Marisa asked, "Can we go home now?"

Sarah said, "Sure. Are you okay?"

Marisa said, "Yeah, but I want to go home."

As the five cousins walked out of the park exit, they heard a commotion. Christian looked back and saw a police car and several people near the ice cream truck.

He said to Sarah, "Stay here with the girls. I want to see if someone needs my help."

Sarah said, "Okay, but be careful." She moved the three girls to the fence.

As Christian walked closer, he could see a young man in handcuffs. Several people were standing around. He asked a young woman, "What's going on?"

The woman replied, "I was out jogging in the park, and I saw that man with a gun. He tried to rob the ice cream vendor, but those two guys jumped him. It was so scary."

Christian asked, "Was anyone hurt?"

The woman said, "No, thank God. I'm glad I didn't bring my daughter here today. We usually get ice cream, and this guy had a gun."

Christian walked back to where Sarah and his three little cousins were sitting on the ground.

Marisa asked, "Why are the police here?"

Christian grimaced, half-smiled, and said, "Everything's fine. Someone called the police because they dropped their ice cream cone."

Marisa said, "That's silly."

Christian whispered to Sarah, "There was a guy who tried to rob the ice cream vendor."

Sarah whispered back, "Oh, my God! Was anyone hurt?"

"No, they had the guy in handcuffs. Sarah, he had a gun! If Marisa didn't want to go home, we would've been standing there with our cousins!"

Sarah said, "We have to tell our parents and have them call the girl's parents. They might see it on the news. I'm more afraid now than ever of this invisible force. I thought it was strange that Marisa wanted to go home right away. Maybe she had a premonition of what was about to happen. I won't ask her because I don't want to scare her."

"Yeah. We're going to have to move these mystery hunts along faster so no one gets hurt. I'm not sure the invisible force had anything to do with this, but I don't want to take any chances. We need to be especially careful and look around when we have kids with us."

Sarah had a few tears in her eyes and nodded.

As the cousins walked home, Marisa asked, "Are you on a treasure hunt?"

Sarah answered, "Well, these are mystery hunts, but sometimes we find lost treasures."

Christian said, "We're learning on our mystery hunts. Sarah and I asked you honey bee and honey jokes. Honey bees are really important because they fertilize flowers and trees. **Honey bees drop ___________ onto other plants and flowers, and this**

produces seeds, so the number of plants and flowers grow and multiply. The honey bee, other bees, insects, and birds spread seeds so food can grow. We need to protect honey bees."

Ameliah asked, "What about kids? Are honey bees important for food for kids too?"

Marisa answered, "Of course, kids are people. We eat plants like strawberries and carrots. We need honey bees, birds, and insects, too."

Sarah smiled and said, "Here we are. You can stay at Marisa's house, and your moms will pick you up before dinner."

"I had fun," Sarah said as she and Christian walked home.

"Yeah, but they're a handful. I'd never go to the park with the three of them without you or my parents."

Sarah nodded and said, "Yeah, one person per child."

Christian walked Sarah to her house and then walked home. As he came in the door, his mom said, "Your Uncle Raymond called you and wants you to call him back."

Christian said, "Thanks, Mom." He shared what happened at the park. His mom decided to call the families later in the evening.

Christian called his Uncle Raymond.

"Hello, Christian. Thanks for returning my call," Uncle Raymond said. "There's another missing object."

"Yeah, a musical composition with a harmonic analysis calculation written on the back. Someone took it from the local high school music room."

"I'm not surprised. The four of you are on top of everything. I'll pick you and Sarah up tomorrow at 9:00. I'll drive you to school and introduce you to Miss Bell. But I should let you know I spoke with a detective in Great Britain today. A team of thieves is stealing mathematical and scientific objects in France and England. We may have our suspects."

"Did you say France and England? They sound like the suspects. Do you think they're in the US?"

"I'll know more tomorrow. We're having a video conference call with the detectives on the case. They believe this team of thieves is indeed here."

"Uncle Raymond, I need to share a couple of things with you. We were at the park, and some guy tried to rob the ice cream vendor with a gun."

"Your mom called me, and I checked with the officers. I'm glad none of you were hurt, and I'm glad none of you saw the incident happen. Are you all okay?"

"Yeah. Sarah's a little shaken up. But we're okay."

"I've spoken with my boss, and we're going to have a few officers walk the park during the day when families are there."

"Thanks," Christian said. "One more thing. I know Mr. Bigollo didn't mention this to you because he didn't think it was important. His nephew owns a bookstore in Great Britain and wanted his uncle to give him the two stolen books. Mr. Bigollo wouldn't agree. He doesn't want the books sold because they're family heirlooms. He's tried to reach out to his nephew, but he hasn't returned any emails or calls.

"Also, Mr. Bigollo mentioned his nephew was involved with the wrong crowd several years ago. I asked him if he thought his nephew might have anything to do with the missing books, and he said, 'Absolutely not!' Mr. and Mrs. Bigollo are gifting some of their family heirlooms to his nephew, but his nephew doesn't know it."

"Interesting. I'm going to follow up on this info. Thanks, Christian. Maybe the wrong crowd is this team of international thieves. I hope I'm wrong because this would break Mr. Bigollo's heart!"

"Should we still plan on meeting with Miss Bell?"

"We should proceed with the investigation. My team of detectives will work with the British police. The detective I spoke with mentioned one of the suspects might have a connection with our town, and that could be Mr. Bigollo's nephew."

Christian said, "They must be the suspects! And this must be the connection. I was sure it was someone wanting to keep the objects for himself. I guess I was wrong. I'll call Sarah. We'll be at my house at 8:30."

Uncle Raymond said, "Sounds good, see you tomorrow. Please don't mention anything about Mr. Bigollo's nephew."

◆ ◆ ◆

The next morning Uncle Raymond picked up Christian and Sarah and drove to the high school. They walked into the music room, where Miss Bell was organizing music stands.

"Hello, Miss Bell. I'm Detective Dawson. This is my nephew Christian, and his cousin Sarah."

Miss Bell said, "Hello, Sarah, I just received your phone message this morning, but Detective Dawson told me about your visit today. It's very nice to meet both of you."

Christian asked, "Are you new here, Miss Bell?"

"Yes, thank you for asking. I have a dual major in music and mathematics, and this is my first year teaching. Let me tell you about the stolen artifact. Scientists have authenticated Albert Einstein's handwriting on the back of Bach's symphony. It's a rough mathematical equation similar to *Harmonic Analysis*. Many famous mathematicians were musicians. **Albert Einstein played the violin and the ____________."**

Detective Dawson asked, "Is this musical composition on loan to the school?"

"Yes, from the Einstein Museum. Other pieces on display are musical compositions, a violin, one of the first harmonicas ever designed and constructed, and a Rees harp. **Another name for a harmonica is a French __________."**

Detective Dawson asked, "Do you have any idea who took the Bach composition?"

"I don't. I'm new at the school. We opened the exhibition on Friday at 5:00. We reopened it on Saturday afternoon. I was here, and so were the other teachers. There were so many families at the event. The principal and teachers at the school agreed the exhibition was a splendid way to meet everyone and share my love of music and math."

Sarah asked, "Why was Einstein so interested in Bach?"

"Johann ________________ Bach is recognized as one of the most 'mathematical' composers of all time. I'm sure this is why Einstein was so fascinated by Bach. *Harmonic Analysis* was a new concept when Einstein was a young man."

Christian asked, "Did any of the visitors spend more time looking at the object?"

"Several families stopped to look. But no one was so interested I might consider them a suspect. I feel responsible, and while the school has insurance, this could be a bad mark on my reputation. I hope you can help find the composition and return it to the school so we can return it to the museum."

Detective Dawson said, "This is the second object on loan from the Einstein Museum, so I'm sure they want both objects returned as soon as possible."

"Thank you, Miss Bell. We appreciate your time," Christian said. "I'm sure my Uncle Raymond, I mean Detective Dawson, will update you on the case."

Miss Bell said, "Thank you very much!"

◆ ◆ ◆

The next day Sarah invited her three friends for lunch.

"Hi. Come on in. Olivia and I made macaroni & cheese and fruit salad. She baked chocolate-kiss peanut butter cookies."

"This looks great!" ______________ said.

Christian said, "We met with Miss Bell at school. She has no idea who took the object. Miss Bell told us some interesting facts about Einstein's fascination with Bach. He wrote a rough draft of a theory using mathematical equations similar to *Harmonic Analysis*."

Sarah said, "She's upset over the loss of this important artifact. She's worried it might hurt her reputation."

_____________ said, "It's not her fault. She was just trying to meet the parents and the kids and teach them about famous math and musical objects. No one should blame her."

Sarah said, "I don't think anyone's blaming her. The Einstein Museum loaned this artifact to the school. But this is the second missing object that belongs to them. We're all worried that we won't find these artifacts so they can be returned to the museum."

_______________ asked, "Are we getting any closer to finding everything?"

Christian said, "It takes one break or one good clue to solve a mystery. The police are working on this case too, and they may have a break in the case. There's a team of thieves stealing math and science artifacts in France and England. The detectives in Great Britain believe these suspects are in the US, and one of them has a connection with our town."

_____________ said, "We probably shouldn't get involved with a team of international thieves."

"I agree," Sarah said. "Christian, what does Uncle Raymond think about this team taking the artifacts?"

"He said it's an important lead. But he suggested we keep searching, as long as we're safe."

____________ said, "On that note, this may sound a little strange, but we learned the number seven meant perfection or _____________. What if there will be seven objects or seven cases of missing objects? None of us believe in coincidences. We can't explain all the connections."

_____________ said, "So, if the suspects take seven objects, they already own the framed butterfly, two books, the painting, the photograph, and the musical composition, and they would only take one more object. But if there are seven cases, and we're not counting missing artifacts, then two more objects will go missing."

Christian said, "There are too many things we can't explain. Let's keep researching. We might be able to locate the next missing object. This international ring of thieves could be the suspects, but if they're not involved, then we still have a mystery to hunt for answers."

_______________ said, "We learned Pythagoras believed the number seven was the most spiritual of numbers."

Sarah said, "When we spoke with Prof. S earlier, he quoted Pythagoras on the connection between music and math."

Christian said, "It makes sense there would be a connection between both."

Sarah added, "We should look online for math artifacts on display. We can narrow our search to local objects."

______________ said, "It's always good to get together. Four heads are better than one, and eight eyes are better than two."

The friends laughed.

"We're narrowing our search," ____________ said, "and there were so many people at these events, but most of them were parents and children. The framed butterfly and the books were the only objects not on display and were in private homes."

______________ asked, "Why don't we investigate connections between Mr. Bigollo and Ms. Parvana? They live in different counties and probably don't know each other. We should contact Mr. Bigollo and ask if he has any relatives or friends in this county. If he doesn't, we can rule out any connection."

Christian said, "Great idea! I'll call Mr. Bigollo tomorrow. Let's call it a day."

◆ ◆ ◆

The next day Christian called Mr. Bigollo.

"Hello, Christian, any word on my books?"

"Not yet," Christian said. "Three other objects are also missing, a painting, a framed photograph, and a musical composition. Do you know Ms. Parvana? She's an elementary teacher at our local school."

"No, but my wife's niece attends elementary school in your town."

Christian asked, "What grade's your niece in?"

"She's in the second or third grade. Our niece has challenges, but she's doing very well. Her family visits our home because we board horses, and she loves to ride ponies. We don't visit their home often. My wife hasn't mentioned the name, Ms. Parvana."

Christian said, "Ms. Parvana teaches elementary children, and her framed butterfly was the first object taken."

"Do you believe a child is doing all this?"

Christian said, "No, I don't, but it's a good clue. I'll call my Uncle Raymond and update him. Thank you again."

"Please thank everyone for their help."

BACH & HARMONIC ANALYSIS

Your name: ____________________ Friend's name: ______________________
Your favorite song: __________________ Musical artist's name: ______________

CHAPTER SIX

"TO INFINITY AND BEYOND..."

Two nights before Sarah planned to babysit her three cousins, she awoke from a deep sleep and said, "*To Infinity and beyond* ..." She repeated this over and over ... "*To Infinity and beyond* ..." "*To Infinity and beyond* ...."

Sarah and her three cousins, Marisa, five years old; Anita, three; and Ameliah, two years old, enjoyed a Friday night of pizza and chocolate chip cookies at Sarah's home. Her cousins loved the children's animated movie, *Toy Story*. They ran around the house with their hands up in the air, pretending to fly. They repeated the line from the movie, "*To Infinity and beyond* ..." over and over! Sarah chased them around the house. She was happy to take a break from the mystery hunt and laugh and play with her three 'miniature sweeties.'

She read three books to her cousins, *Blueberry Pi*, *Heart of Gold*, and *Cleopatra Cat*, by K.C.R. Palmer. Before they went to bed, she read them the books *Boomerang* and *Numbers from Heaven* by the same author. Sarah loved to read to her cousins as much as she loved to read books herself before falling asleep at night. Sarah thought for a moment. Was I repeating '*Infinity and beyond* ...' because we planned to watch *Toy Story*?

◆ ◆ ◆

The next morning, Christian came over to Sarah's house. "I'm exhausted!" Sarah said with a sigh as she opened the door. "I love babysitting, but between chasing them around, dinner, snacks, helping them brush their teeth, getting them ready for bed, and reading a few books, I can barely move this morning. Olivia was at the library studying and came home around 9:00 p.m. to help me. I feel like I'm thirty!"

Christian laughed and said sarcastically, "Yeah, at age thirty, you can barely walk. But I'm exhausted just listening to you!"

Sarah asked, "Did you find any new information online?"

"Not yet, but I'm still looking." Christian paused and asked, "Let me guess, *Toy Story* again?"

"Yep, in the last two years, we watched *101 Dalmatians* 1001 times! *Toy Story's* a good break. You're right. We need at least three people to babysit these miniature sweeties, so I'm counting on you and Olivia to help next time."

"Oh, no! The park was enough for me! It takes a special gift to babysit our cousins. I'm sure Olivia will help you."

Sarah rolled her eyes and smiled.

Sarah asked, "Did you hear from ______________ and ______________?"

"Yep, they're searching online every day, but nothing new. I feel sorry for Miss Bell."

"I know, but we'll find the objects. We solved the last mystery. It just takes time."

"Yeah," Christian said with a half-smile, "*To infinity and beyond ....*"

◆ ◆ ◆

______________ and ______________ were spending the day at the water park in town.

______________ said, "It has to be 90 degrees today, but it feels awesome anyways."

______________ said, "This is great! There are so many people at the water park. After this, let's buy hamburgers, fries, and blueberry slushies."

"Sounds good. Christian asked us to come to Sarah's house tomorrow at 10:00."

Sarah and Christian spent the day mapping out each object's owners and the original locations of the missing artifacts.

Sarah said, "Only one owner lives in another county."

"Two of the objects were on loan from the Einstein Museum, in ____________, but this shouldn't be part of the equation."

"I agree."

Christian said, "____________ and ____________ are meeting us tomorrow. They're at the water park today."

"I thought about going, but I need to spend the day with you to talk about this mystery hunt. We'll go this weekend. It's going to be over 90 degrees again on Saturday. It's so nice of Uncle Raymond to buy us season passes."

"Yeah. Uncle Raymond said it was the least he could do for all our help. It's still early in the summer. We have plenty of time to use our season passes."

Sarah asked, "Any new info pointing to where the suspects may grab the next object?"

"We've checked online, and nothing stands out."

Sarah said, "If ____________ is right, then the suspects will only take two more objects, making seven individual cases."

"Well, until we see more math objects in the newspaper or online, we can't guess what's next. Not unless this ring of thieves from Great Britain stole several things before they came to the US. Then we're way past the number seven. I'm going home. I'm helping my mom. I'll see you tomorrow."

Sarah asked, "What chores are you doing?"

Christian half-smiled and said, "My mom tells everyone in our family my dad and I are afraid of the dishwasher. I'm cleaning my bedroom and doing dishes."

Sarah smiled and asked, "Why are you and your dad afraid of the dishwasher?"

"It's a family joke. My dad and I never open the dishwasher to put in the dirty dishes. We put them in the sink. My mom tells everyone we're afraid of the dishwasher. In reality, she's sure we won't open the dishwasher for fear it's filled with clean dishes, and we'd have to put them away."

Sarah smiled and said, "Hysterical!"

"Yeah," Christian said with a half-smile. "She also says we have refrigerator blindness because we can't find anything in the fridge, and she has to find everything."

◆ ◆ ◆

The next day the friends met at Sarah's house.

Sarah asked Christian, "Did the dishwasher bite you, or did you have to fight it off?"

Christian half-smiled and rolled his eyes.

_____________ and ______________ looked at both of them and shrugged.

_____________ said, "I found something, but it's already missing."

_____________ asked, "What is it?"

_____________ said, "It's the *Infinity* symbol with stars. An artist created a rendering of the symbol in metal, and there's also a painting of it. It was displayed at the Memorial Art Gallery."

Sarah asked, "How large is the sculpture?"

"From the online article, it looks around the size of a loaf of bread. It sat on a pedestal. The artist is a famous sculptor. I printed out the picture for you."

Christian said, "I'm meeting with my Uncle Raymond tomorrow. I called Mr. Bigollo. He doesn't know Ms. Parvana, but his wife's niece goes to the school where Ms. Parvana teaches. My uncle and I'll talk about the evidence. I hope he updates me on the ring of thieves from, as my mother describes it, 'across the pond.'"

______________ said, "Each time something went missing, families visited these buildings. A kid probably took the objects."

Christian said, "I don't think so. It must be an adult. The suspect was in Ms. Parvana's home, and then Mr. Bigollo's home in another county."

____________ asked, "What if it's one of the owners of an object, and they're trying to throw us off by pretending to be a victim?"

Sarah answered, "Yeah, they might be suspects. They each love math objects. But I'm sure the police are investigating any connections. Christian, is that the reason you're meeting with your Uncle Raymond?"

"Yeah," Christian said. "He won't be able to tell us much. I'm just hoping we can brainstorm. I'll ask Uncle Raymond to introduce us to someone in charge of the art gallery."

"Sounds good," ______________ said. "You know, it could be a couple of older teens? Anyway, let us know if we can help you."

The next day, Christian waited an hour for his uncle.

Uncle Raymond knocked on the door and walked in. "Sorry, I'm late. I was going to call, but I had it out with my new boss. He wants this case closed and whoever is responsible for stealing these objects to go to jail. He has a lot of pressure on him to close all our cases ASAP."

"I'm sorry. That's a lot of pressure on both of you. Thanks for meeting with me. I know how busy you are. Sarah and our two friends wanted me to thank you for the season passes."

Uncle Raymond said, "I'm proud of each of you. I looked into Mr. Bigollo's niece attending our elementary school. Her name is Mary Cartwright, and she's one of Ms. Parvana's students. My detectives agree there are no criminal connections. So, it might be a coincidence. And the Criminal Investigation Department or CID detective in Great Britain believes there is a connection to our area. He wouldn't provide me with any other information for now."

"Uncle Raymond, Sarah, and I met Mary and her friend Vanessa. We understand you can't give us much information, but did your investigators find other connections between the owners of the objects?"

"We completed part of the investigation. While I can't go into any details, we can't find any connections."

"Did you consider one owner of a missing item is taking these objects?"

"Our investigators agree there are no red flags. But we'll continue to look at that possibility."

"Would you introduce us to someone at the art gallery?"

"I spoke to the Assistant Director. His name is Aseem Apar. Here's his phone number. You can call him and set up a meeting."

"Thanks, Uncle Raymond."

"Anything for you, buddy." Uncle Raymond paused and said, "Why don't I call? I can pick you and Sarah up again, and the three of us can ride to the gallery."

◆ ◆ ◆

The next day Uncle Raymond, Christian, and Sarah walked into the art gallery to meet with Mr. Apar.

"Hello, Aseem. This is my nephew, Christian, and his cousin Sarah. I spoke with you about their team solving mysteries."

"Hello, Detective Dawson. It's nice to see you again. I just called your station. I was on my way to our meeting, and two cars hit me at once. Thankfully, no one was injured. There was a major accident somewhere in town, so the police couldn't send someone right away. But none of us were hurt. We took each other's insurance information and license plate numbers, and then took photos with our cellphones."

Uncle Raymond said, "Are you sure you're okay? Maybe I should drive you to the hospital?"

"No, thank you. It was a fender-bender for all of us, but the strange thing is, the other two drivers said they saw a shadow of some sort and tried to break to miss the shadow. I didn't see the person or the shadow as they described it. But their brakes weren't working, and they hit me. I could drive my car, but the other two drivers had their cars towed. It was very strange. I've never been in an accident before this one."

Uncle Raymond, Sarah, and Christian looked at each other. Sarah whispered to Christian, "The invisible force? Maybe it's not so invisible!"

Christian whispered back, "You're right. We have to move this mystery hunt along, or someone's going to get hurt."

"Please follow me to the room where we displayed the sculpture."

Christian said, "Thank you for meeting with us, Mr. Apar. We're glad you're okay. We toured here on a school field trip. It's an amazing art gallery."

Sarah asked, "How long have you worked here?"

"This is my seventh year as the assistant director. Our director is out of the country on an international tour of famous art galleries. Our organization is both a museum and an art gallery. We sell artwork for local artists. This situation is very upsetting."

Christian asked, "Mr. Apar, was the gallery opened when you realized the object was missing?"

"Yes."

Christian asked, "Was the sculpture in a locked case, or was there an alarm attached?"

"No, it was on a pedestal. It wasn't in a locked display. We attach alarms and locks to several of the paintings and sculptures. But I decided to unlock the displays and disarm the alarms. There were so many children running and playing. I didn't want to frighten them if they accidentally triggered an alarm."

Sarah asked, "Was the sculpture worth something?"

"The current owner purchased it for $5000. It's on loan to our gallery. The sculpture is a visual example of the mathematical concept of infinity. The sculptor V.M. Ross, Jr. named his sculpture, *To Infinity and Beyond to the Stars."*

Christian asked, "Do you have a camera security system?"

"It was such a busy day with so many children; we turned off the cameras. I don't like filming families. It was my decision and one I now regret. Our gallery is insured, but my concern is finding the sculpture and returning it to the owner."

Uncle Raymond shook Mr. Apar's hand and said, "Thank you for your time and for answering our questions. Please don't question your decision to turn off the alarms and the cameras. I agree with both the decisions and when your director returns in the next few weeks, I'll contact him. As we discussed yesterday, we're investigating other missing math objects."

◆ ◆ ◆

As Uncle Raymond drove Christian and Sarah home, he said, "I'm so proud of both of you. Those were insightful questions. I hope one or both of you consider a career in law enforcement." He chuckled and said, "Or maybe you could start a detective agency to find missing objects taken by paranormal invisible men."

Christian and Sarah laughed out loud.

"But seriously, if what Tata described is an invisible force trying to stop you, you all have to be careful. I'm not sure what this is, but the whole thing's a little strange. It could've been a reflection. Strangely, the other two drivers' brakes failed. I'll look into it."

Christian and Sarah looked at each other and nodded.

Sarah said, "Uncle Raymond, this past weekend, I was babysitting my three little cousins, and we were watching *Toy Story*. We were running around all night saying, '*To Infinity and beyond ....*'"

Uncle Raymond said, "*Toy Story* is one of my favorite movies. And after spending time with your team, and my years of experience as a detective, I agree, there are connections."

Christian smiled and asked, "Uncle Raymond, don't you and Aunt Paris watch *Toy Story* together?"

Uncle Raymond smiled and said, "I first saw *Toy Story* at the movie theater with two of my nephews, but I may own a copy."

The two cousins laughed out loud, and Christian said to Sarah, "I told you, he's a softie when he's not working."

Uncle Raymond smiled.

◆ ◆ ◆

The next day the friends met at Sarah's house.

Christian said, "We met with the art gallery assistant director yesterday. They turned off the alarms and cameras because of the number of kids visiting the gallery."

______________ said, "That's too bad! An actual video of the suspect taking the sculpture would solve the case. And I could see a ring of professional thieves stealing a sculpture at an art museum, just like in the movies!"

___________ asked, "What's *Infinity* again?"

___________ said, "I researched the word *infinity* on the computer, and hundreds of hits popped up on the search engine. I narrowed my Internet search to the words *infinity* and math."

Sarah asked, "What did you find?"

"The symbol reminds me of the number eight," ____________ said. "I'll read the information on one science site. '**Although *infinity* is larger than any number in existence, it's not a real _________. If you add *infinity* to *infinity*, you'll realize *infinity* + *infinity* = _______.** *Infinity* is not only enormous, but it's also endless.'"

Christian asked, "But doesn't *infinity* depend on time and space?"

"Wow! That's a tough question!" Sarah thought for a moment and asked, "If there was no time or space, is that *infinity*? If someone is outside time and space, is *infinity* missing?"

_____________ and _____________ looked puzzled.

_____________ said, "We should focus on the math of *infinity*. Those questions are too tough for us to solve today, and tomorrow, and '*To Infinity and beyond ....*'"

Christian said, "Funny."

"So, looking at the symbol, it has no beginning or end," ______________ said. "It's the perfect symbol for infinity."

Christian asked, "Would you keep reading?"

"The *infinity* symbol is called the _______________ meaning ribbon in Latin. The Latin word infinitas or *infinity* means _______________. A clergyman and mathematician ___________Wallis created the symbol in 1655."

_______________ asked, "Christian and Sarah, how can someone or something be outside space and time?"

Sarah answered, "We go to Bible Study class together. God created space and time. I have the note I wrote down in my laptop case. 'God is not restricted by physical elements or conditions. If God is outside space and time, then *infinity* couldn't affect Him, and He wouldn't have the constraints of space or time.'"

_____________ said with a smile, "This is way too deep for me. But it's something to think of before I go to bed. This should put me to sleep."

The friends smiled.

Sarah said, "This object is the ultimate symbol of math. It's a question no one may ever answer."

_____________ said, "I wonder if this will be the last object missing, the seventh object."

"Time will tell ... get it? We should take a break," Christian said.

Sarah suggested, "Why don't we take a few days off from the mystery hunt?"

_____________ said, "I agree. It's our summer vacation. Why don't we go to the water park on Saturday?"

Sarah said, "The water park sounds great! We can meet at 11:00. Don't forget your sunscreen."

Christian said sarcastically, "Yes, Mother."

Sarah rolled her eyes and said, "You'll thank me when you don't burn."

◆ ◆ ◆

The friends met at the water park on Saturday. Parents and children were swimming, riding on the water slides, drinking slushies, and eating ice cream.

Sarah reminded her friends to put on their sunscreen. They spent the entire day swimming and sliding down the tall water slide.

_____________ said, "We learned a lot about the number seven. And since the *infinity* symbol resembles the number 8, I researched the meaning of the number eight."

_____________ asked, "What does it mean?"

_____________ said, "It can mean *harmony* and a new beginning. It gives me the chills! We were discussing music and *Harmonic Analysis*. When we finish this mystery hunt, we'll move to a new beginning, the number eight, one symbol for *infinity*."

_____________ said, "Okay, we can't leave this mystery behind."

Christian said, "Sarah and I agree we should try to solve these mysteries as quickly as possible. Once we find out why someone is taking the objects, we can narrow the search for the suspects. We're learning a lot about mathematics and science.

Remember, we learned the mathematical symbol *Pi* goes on and on ... *To Infinity and beyond ....*"

Sarah said, "Okay, enough for today. Does anyone want a fruit slushie?"

_______________ said, "Yeah, thanks. Strawberry slushies for the two of us."

Christian said, "Do they have just plain vanilla?"

Sarah smiled. "No, but just plain water is close to just plain vanilla. Do you want ice water?"

Christian said, "Funny. Here's a twenty, please buy me an orange slushie. The drinks are on me."

Sarah asked, "An orange slushie? Walking on the wild side?"

Sarah and _____________ walked to the concession stand.

Ms. Parvana walked over to the friends, and said, "Hello."

Christian said, "Hello, Ms. Parvana. How are you?"

"I'm well. How's the case going? I read about an object missing from the art museum."

"Yes, a metal sculpture of the *infinity* symbol. Would you like to sit with us?"

"I can only sit for a few minutes. I was just on my way home."

______________ asked Ms. Parvana, "Can you explain *infinity*?"

Just then, Sarah and ____________ walked over with the four slushies.

"Yes, I can. Hello, Sarah. How are you?"

"Wonderful, Ms. Parvana! These are two of our best friends, ____________ and ____________. We bought slushies. I'll run over and buy you one. The drinks are on Christian."

Ms. Parvana smiled and said, "Well, that'll be messy!"

Everyone smiled.

"No, thank you. I'm all set. Christian asked me to explain *infinity.* There's a funny example used over and over called the *Monkey Theorem.* Let's suppose we put a monkey in a room with a typewriter. **If the monkey had an infinite amount of time, could he write ___________________'s *Hamlet*?** It's an excellent example because it's impossible, similar to *infinity*."

_____________ laughed, "It could be possible if it was a very smart monkey."

_____________'s three friends smiled and rolled their eyes.

Ms. Parvana asked, "Does infinity have different sizes?"

The friends looked at each other.

Sarah said, "Well if there are smaller numbers such as one and two, and larger numbers such as 1000 or 2000, infinity has numbers of different sizes."

"Excellent!" Ms. Parvana said. "Another question, what is the sum of *infinity* + 1?"

The friends smiled and looked at each other again.

______________ asked, **"Is the answer __________?** We cheated a little because we read the answer online today."

"Excellent! And it's never cheating to read and study mathematics. Well, I'm going home. Please keep me up to date when you're closer to the answer to our great mystery, 'Who loves math enough to steal *mathematics*?'"

The friends decided they had enough fun in the sun. Olivia came to pick up Sarah and her friends.

Olivia asked, "How was the water park? I hope you didn't get sunburned."

Sarah replied, "We each put on our sunscreen at the park."

Olivia laughed and said, "I'm sure my sister insisted on sunscreen for everyone!"

The friends laughed except for Sarah, who smiled and rolled her eyes.

When Christian arrived home, his mom asked, "Did you have a nice time?"

Christian answered, "It was a perfect day. We met Ms. Parvana at the park, and she taught us the math concept of *infinity*."

His mom laughed, "You can't go anywhere without these mysteries following you. Did you remember to put on sunscreen?"

Christian half-smiled, rolled his eyes, and said, "Yes."

"Oh, I forgot," Christian's mom said with a smile, "Sarah was with you. No worries. Sweetie, you received an envelope in the mail."

Christian said, "Thanks, Mom."

"TO INFINITY AND BEYOND TO THE STARS"

CHAPTER SEVEN

SYMMETRY IN FLIGHT

Symmetry in flight ... honey bees and hummingbirds, a punctuated sigh, one vibrant autumn leaf ... falling, falling, falling ... Symmetry in flight. Sarah wrote her thoughts in her journal every week.

Sarah and Christian walked to the park and flew their kites. It was a beautiful summer day with a strong breeze. Sarah brought her butterfly kite, and Christian unveiled his new kite, a golden honeycomb with four honey bees, one at each corner.

Sarah smiled and said, "I guess those honey bee jokes inspired you to build a new kite."

"Yeah," Christian said. "Honeycombs and honey bees fascinate me. The same pattern repeated over and over. It's so cool! My dad and I finished it last night. He calls it wallpaper symmetry. It wasn't easy to make. We used a compass and a ruler. Dad mentioned most people use both tools to create this pattern. What's amazing is that honey bees create this pattern without a compass and ruler!"

Sarah smiled, "A honey bee can't hold a ruler and compass."

"A honey bee could hold a compass and ruler if the instruments were very, very small," Christian said with a half-smile.

Sarah smiled and asked, "What's wallpaper symmetry?"

"It's when a pattern repeats until it covers a specific area. People create these patterns in art and architecture. It took us four hours to map out and paint these combs."

"It's beautiful!"

"Thanks, I'll tell my dad. He worked for hours creating the outline. We worked on new kites for our three exhausting cousins."

"They'll be so excited!"

"They love to giggle. It's a surprise, but my dad's recording each of their giggles, and he'll add a small recorder to the kites. When each one flies her kite, she'll hear her giggles coming out of the voice recorder."

"They'll love it! You should give the kites to the girls when you're babysitting them with me."

Christian half-smiled and said, "Nice try, Sarah."

"Do you think you and your dad could make kites for Vanessa, Mary, and Kennedy?"

"That's a fantastic idea! They're sweet and love math and butterflies. So, what if we build a unique butterfly kite for each of them? I'll ask my dad tonight and let you know."

Sarah asked, "What if I ask Ms. Parvana to record her two students' giggles so you can attach the recorder to their kites?"

"If Ms. Parvana will record their giggles, you can drop off the small recorders and instructions at her home. I'll give Aunt Toni a recorder to record Kennedy's giggles."

"Thanks."

◆ ◆ ◆

Christian received a text from Uncle Raymond about another missing object, a large stained-glass rendering of a butterfly! Christian said aloud, "It can't be!"

It was stolen from the art museum. Assistant Director Apar contacted his uncle to make a police report.

Christian called his Uncle Raymond. "Another butterfly? Really?"

Uncle Raymond said, "Yes, this mystery hunt has come full circle. Another butterfly! Director Apar emailed me a photo of the stained-glass butterfly. It's amazing! I'm meeting with him this afternoon to take the report. I just sent the photo to your cellphone. It's titled *Symmetry in Flight.*"

Christian heard a ding on his cellphone, and the photo of the stained-glass butterfly popped up on his screen. It was amazing!

"We've already spoken to Asst. Director Apar. Would you let him know we'll add this to our mystery hunt? I'll let Sarah, _____________, and __________ know."

Uncle Raymond said, "I'll speak to Aseem and let him know. Thanks. If I hear about any other information I can share, I'll text you. You won't have to meet with him again. You've already toured the art gallery."

"Thanks, Uncle Raymond."

Christian texted Sarah and their two friends and sent the photo of the stained-glass butterfly. They each had the same reaction, "What?"

Sarah called Christian. "Really? A butterfly. It's gorgeous, but this is now seven cases with eight missing objects. What does Uncle Raymond think?"

"He doesn't believe we need to meet with Mr. Apar because we already toured the art museum. He'll let us know if there's any additional information he can share."

Sarah said, "Well, there's not much we can do. If this ring of thieves is here, Uncle Raymond will have to deal with them. Does the object have a math connection?"

"The object is titled, *Symmetry in Flight*, and I know symmetry is a math concept. So, the missing object is connected to the other math objects."

Sarah said, "Let me call you right back."

Several minutes later, Christian's phone rang.

Sarah said, "I looked up symmetry and butterflies. A butterfly is considered to have bilateral symmetry, with one side of the butterfly being a mirror image of the other side. In mathematics, symmetry includes translation, reflection, and rotation. So, this butterfly is a perfect example of symmetry in mathematics."

Christian said, "So, you're right. The same suspects must have taken the object. We have eight missing objects, but seven cases. We shouldn't change our plans to meet in the park. I have all the kites. I'll see you in a few minutes."

◆ ◆ ◆

Sarah and Christian flew their kites and had ice cream before leaving the park. There was a line at the ice cream truck.

"It's a long line to wait for just plain vanilla," Sarah said with a smile.

Christian smiled and replied, "I hope the kids standing behind us aren't upset because you're taking so much time to order your chocolate ice cream, with sprinkles, nuts, whipped cream, and a cherry on top?"

"Okay, you made your point."

Sarah said, "I'm glad we told Uncle Raymond about the attempted robbery here at the park. I see police walking around here all the time now."

"Yeah. Uncle Raymond said he would have a 'police presence' in the park now because of all the families hanging out here in the summer. Do you feel safer now?"

Sarah said, "I do. But only here. I'm still worried about the invisible force, but I feel much safer bringing our cousins here, now."

◆ ◆ ◆

The friends spent the day at Christian's home. They sat at the dining room table with open laptops.

Christian asked, "So, what do you think about the stained-glass butterfly being stolen?"

_______________ said, "We both think the object is incredible. And we agree it was taken by the same suspects."

Christian said, "My Uncle Raymond will call us if there's any more information. What did the two of you do yesterday? We went to the park and flew kites."

_____________ said, "I finished cleaning my room and mowing the lawn, so my weekend's free."

____________ said, "I did the same. We planned on going to the water park this weekend, so we finished our chores. Why don't you come with us?"

Sarah said, "It sounds great! Thanks."

Just then, Christian's mom came into the room with a bowl of fresh fruit, donuts, and bagels.

The friends said, "Thank you."

"Christian, your mail's still on the table. Here it is. There's no return address on the envelope."

Christian said, "Thanks, Mom. I forgot about the envelope." He looked over the front and the back of the large manila envelope. His mom was right; there was no return address. Christian opened the envelope. Inside was a picture of the framed butterfly and a typewritten letter.

The three friends saw the butterfly photo and looked at each other.

_____________ asked, "What does the letter say?"

Christian looked up after he read the letter and said, "This might be from the person who took the objects. I'll have to call my uncle right now. There might be fingerprints or other evidence."

Christian called his Uncle Raymond.

Uncle Raymond said, "Don't touch the envelope, picture, or the note. I'll be right over with an evidence bag. Was the note threatening in any way?"

"No, Uncle Raymond. It's a nice poem."

"No one touches the photo, envelope, or note. My uncle is coming over with an evidence bag." Christian took a photo of the letter.

The friends stepped away from the dining room table and sat at the kitchen table.

Christian walked into the laundry room to speak to his mom. She walked over to the kitchen table and spoke to his friends. "Christian did the right thing. I read the letter, and from its content, you shouldn't be worried. No one signed the letter, but the poem explains why this person took the objects. Are any of you upset?"

_____________ said, "Are you kidding? This is so cool!"

_____________ said, "I agree, it's very cool."

Christian snapped a cellphone photo of the butterfly picture, the poem, and the envelope.

There was a knock on the door. Uncle Raymond and Christian's mom walked into the kitchen.

Christian said, "Hello, Uncle Raymond. Thanks for coming right over."

Uncle Raymond said, "You did the right thing by calling me."

Christian said, "I took a photo of the evidence. I'll read the poem."

SYMMETRY IN FLIGHT

The butterfly brings hope and light
The butterfly is *Symmetry in Flight.*

CHAOS & THE MISSING BUTTERFLY

Fibonacci, Phi, the Golden Ratio, and Pi
Beautiful spirals and Golden stars in the sky.

The Bible, Elements, Liber Abaci, books for the soul
Has God handwritten our future on one *Infinity* Scroll?

Children holding ice cream cones, giggling, and flying kites
Dancing in the sky with stars & numbers in the moonlight.

Lorenz, Pythagoras, Newton, the Big Bang, and Einstein
Did our world begin with a Whisper outside of space and time?

Can we count the stars & numbers to *Infinity*?
Does each life display superlative *Symmetry*?

Why do we desire these magnificent mathematical things?
It's our God-given right to dream and play violins with *Superstrings*.

Uncle Raymond, Christian's mom, and the four friends looked at each other and were quiet for a moment.

Sarah said, "The poem's title is the same as the title for the missing stained-glass butterfly! And I wrote my thoughts in my journal today, and used the words *Symmetry in flight*!"

Everyone was quiet and looked at each other.

Uncle Raymond asked, "Where's the evidence?"

Christian said, "In the dining room."

The four friends, and Christian's mom, followed Uncle Raymond into the dining room.

Uncle Raymond took photos of everything, put gloves on, and placed the envelope's contents into an evidence bag. He then sealed the bag and wrote his name and date along the sealed flap of the envelope.

They went back into the kitchen.

"Thank you. Christian, I'll need your and your mom's fingerprints. You both touched the envelope, so we can rule out your fingerprints. I'll deliver the evidence to our forensic team. But the four of you will figure out what the poem means before our team does ... if history repeats itself," Uncle Raymond said with an enormous grin.

The friends smiled.

Christian's mom asked, "Raymond, do you think they're in danger?"

"No," Uncle Raymond said. "This sounds as if it's an apology letter and an explanation. The writer wants Christian and his friends to know he or she loves math and took these items for a personal reason. But how does the suspect know Christian and where he lives? How does the writer know Christian and Sarah fly kites, or is the suspect just writing about what kids like to do?"

"Wait!" Sarah said. "The photo of the butterfly is a photo of the missing framed butterfly."

"Good eye, Sarah," Uncle Raymond said. "Our forensic team will assess if the photo's a copy from the newspaper or a photo of the original framed butterfly. If this photo is of the original, it confirms the person writing the poem is the suspect.

"Before I leave, I need to update you. This is a strange one. British Intelligence has arrested the ring of thieves. BUT the connection is a young man who's related to Mr. Bigollo. He's not one of the ring members but knew them. He turned them into British Intelligence and the FBI. They wanted to steal Mr. Bigollo's books, but they were stolen before these suspects could fly to the US. So, they had nothing to do with these missing artifacts."

Uncle Raymond said goodbye and left for the precinct.

Christian said, "That must be Mr. Bigollo's nephew. The one he's leaving some of his books and family heirlooms to in his will. When I read the letter, I was sure it wasn't this ring of thieves. They wouldn't know anything about me or our team."

Sarah said, "Uncle Raymond's right. We need to put our evidence together. The first thing we need to research is the math concept of *Symmetry*. Then we can unravel the clues in the poem."

Christian nodded.

____________ said, "I agree. We're aware of the other concepts, but he wrote on *Symmetry*, so I'll research it online."

Sarah asked, "Why don't we study this at home? We can combine our research on *Symmetry*. We can meet back at my house on Monday at noon."

Christian said, "I'll send you the photo of the poem and picture to your cell phones so that you can look at the evidence."

The three friends walked home to research *Symmetry*.

◆ ◆ ◆

The friends stopped by and had lunch at Sarah's. Olivia made tacos, homemade salsa, salad, and Mexican Hot Chocolate Cookies.

___________ said, "Please thank Olivia for this great lunch!"

___________ said, "These chocolate cookies are amazing! They have a little kick to them."

Sarah said, "They have a pinch of cayenne pepper."

Christian said, "Sarah, you know I only eat sugar cookies, but these are delicious!"

"Please, someone write down the date and time," Sarah said with a smile. "History is being made! Christian likes something besides just-plain-vanilla or just-plain-sugar-cookies."

They smiled, except for Christian. He half-smiled.

_____________ said, "I'll read a few facts on the *Theory of Symmetry*. '**The line of *Symmetry* or ____________ *Symmetry* divides a specific shape into two identical parts or a reflection of one side.** If we fold a photo of a butterfly in half, one side resembles the other side. If I took a photo of your face and folded the photo in half, one side has half of your chin, half of your lips, part of your nose, one eye, one eyebrow, and one ear. One side of your face reflects the other side.'"

"I researched *Symmetry* last night. **Honeycombs have a ____________ *symmetry*,"** Christian said. **"**My dad and I noticed the symmetry of each of the circles."

Sarah asked, "When you were researching pictures of honeycombs, did you see any honey bees with a compass and ruler?"

Christian answered, "Hilarious Sarah!" Christian looked at ___________ and ____________, and said, "Another inside joke."

"When a shape is rotated _____° it looks the same, even when it's turned upside down," _____________ said, "like a deck of cards. I read something off the Internet, '*Symmetry* in mathematics occurs not only in geometry but in other branches of mathematics. ______________ **symmetry occurs when every point of an image has a matching point the same distance from the central point but in the opposite direction.**'"

"*Symmetry* is a beautiful word," Sarah said. "It reminds me of harmony, reflection, balance, and the patterns on butterflies."

___________ asked, "So, a line of *symmetry* is an imaginary line dividing a picture or object in half, but the sides match?"

Christian said, "Yeah. **There's a concept called _____________ *symmetry*.** If an object is enlarged or reduced in size, the new object has the same characteristics as it did before it was changed."

"I've seen so many things in nature having *symmetry*, sunflowers, seashells, and more," Sarah said. "They have photographs of so many things online. *Symmetry* is beautiful!"

Christian said, "We should call it a day. My Uncle Raymond hasn't called about the letter or envelope. My fingerprints and my mom's fingerprints will be on the envelope. The postal workers left fingerprints by handling the envelope. The suspect who mailed the envelope didn't add a return address or sign the poem. So, I'm sure he or she was careful not to leave evidence."

___________ said, "Probably. But the suspect would need his or her fingerprints on file to match the fingerprints on the envelope. There's not much chance the person was arrested before if it's not the international ring of thieves."

___________ said, "The suspect is trying to explain why he or she took the objects. I'm not comfortable calling this person a thief, because of the poem. Why don't we meet in a few days and try to decode the poem?"

___________ said, "Great idea! I read and reread the poem last night."

The friends smiled and nodded in agreement.

Sarah asked, "Are we still going to the water park tomorrow?"

The three friends nodded. "Absolutely!"

◆ ◆ ◆

The next morning, the four friends met at the water park.

Sarah asked, "Did everyone put on their sunscreen?"

"Yes, Mom, we remembered," Christian said with a half-smile. __________ and __________ smiled too.

The friends enjoyed a day of swimming. They slid down the water slide over and over.

As the four sat on lounge chairs, they heard a soft voice. Kennedy and the two little girls, Vanessa and Mary, peered at the group.

Kennedy asked, "What are you doing here?"

"Hi sweetie," Sarah said. "We came to cool off from a hot summer day. Who's with you?"

Kennedy said, “My mom brought us. Alex races tonight. The three of us are having so much fun.”

Christian asked, “Hello, Vanessa and Mary, do you remember me?”

Mary nodded.

Vanessa said, “Yep, I remember you. I counted 14 umbrellas.”

Sarah said, “How wonderful! I know you and Mary love math. Do you count things every day?”

“Yep,” Vanessa said. “When I’m sad, I count balloons, birds, butterflies, and stars.”

“I’m so glad counting makes you happy,” Christian said. “Kennedy, these are our two friends, __________ and __________.”

The friends smiled and talked with the three girls for a few minutes.

“Well,” Kennedy said, “I’ll walk them back to my mom. Sarah, are you and Christian going to the park this week? My mom said Christian has a surprise for me.”

“Yes,” Christian said. “Why don’t you bring Mary and Vanessa to the park this Wednesday? I’ll bring a surprise for them too. Sarah and I’ll meet you at 11:00.”

“Great!” Kennedy said. “I’ll tell my mom. See you later. Nice to meet both of you.”

Sarah said, “They’re sweet girls. Christian, are the kites ready for the girls?”

“Yeah,” Christian said. “My dad and I built six kites with recorders. I have the three kites for our little cousins, and Ms. Parvana recorded Vanessa and Mary giggling and dropped the recorders off at my house. Kennedy’s mom recorded her giggles.”

Sarah asked, “Are you ready to go home? I’ll call Olivia to pick us up.”

◆ ◆ ◆

On Monday, the four friends met at Christian’s house. They began deciphering the poem.

______________ said, “Let’s begin with the first two lines. The first missing object was the framed butterfly. I’ll call the suspect ‘he or him’ because it’s easier. So, when he writes the stanza on *The Butterfly Effect,* he didn’t use the word *Chaos* or *Chaos Theory*. He has a positive view of butterflies and *The Butterfly Effect.* He uses the word *harmony* for *Harmonic Analysis.”*

Sarah said, “I agree. And the title of the poem is *Symmetry in Flight*, the same title as the stained-glass butterfly taken from the art museum.”

Christian added, “He included each math theory in the poem. He likes children and wants them to play and be happy.”

__________ said, "He may have faith in God, or he's questioning God."

Christian said, "I reread the lines, '*Can we count the stars & numbers to Infinity? Does each life display superlative Symmetry***?** *Did our world begin with a Whisper outside of space and time?'* It's as if he knew Sarah and I were asking questions about *Infinity* and our discussion on God being outside space and time."

Sarah said, "I thought the same thing."

___________ said, "The last line is confusing. The suspect is saying we all have a *God-given right to dream*. He writes about children giggling and playing in the poem."

"I spoke with my Uncle Raymond. There's no physical evidence on the envelope or the papers or any other fingerprints other than the ones they've confirmed. But the photo is of the original framed butterfly," Christian said. "I have an idea, but I need to check into a few things. Why don't you both come to the park on Wednesday?"

___________ said, "Sounds good. Should we bring our kites?"

Sarah said, "Yeah. See you both on Wednesday at the park."

The two friends left.

Christian said, "I need to talk to you about our dreams."

Sarah said, "I know. Either the suspect's heard of our dreams of the stars or numbers, or our dreams are pointing us to the suspect."

Christian paused a few minutes and said, "I have an idea about the suspect."

Sarah asked, "What? Who?"

Christian said, "I don't want to share what I'm thinking, in case I'm wrong. We shouldn't accuse someone unless we're sure. I'll tell you more when we meet on Wednesday at the park."

Sarah said, "But I can't wait! It's one of the owners of the objects, isn't it?"

Christian said, "I need to look into a few things. We can't blame someone for something he or she didn't do unless we're sure. I won't share this with Uncle Raymond, or my family at this point until I'm at least 90% positive."

Sarah said, "Yeah, it's only fair."

"I'll meet you at the park on Wednesday with the new kites."

"Okay," Sarah said. "You shouldn't accuse anyone unless you're sure."

Christian said, "If it's the person I suspect, he or she took the objects for a good reason. So, I want to give the person the opportunity to give everything back."

Sarah smiled and hugged Christian.

Sarah said, "You're always thoughtful. Another reason why you're my best friend."

Christian said, "Treat others as you want them to treat ______."

◆◆◆

Wednesday was a lovely, warm day with a strong breeze. It was 75 degrees and a perfect day for kite flying.

Kennedy, Vanessa, and Mary ran over to the four friends in the park.

Kennedy asked, "Cousin Christian, did you bring our surprises?"

Christian smiled and said, "I did."

He unpacked three unique, beautiful butterfly kites. Each kite had the colors of the rainbow.

Vanessa and Mary giggled and clapped.

Christian said, "These are special kites. Listen." Christian opened the kites and hit a switch on each kite.

Kennedy asked, "Do our kites giggle? We can hear our giggles!"

Christian said, "Yep. Your mom recorded your giggles, and Ms. Parvana recorded the giggles of Vanessa and Mary."

As Kennedy, Vanessa, and Mary flew their kites, the sound of their unique and sweet giggles surrounded the park.

The three girls ran back to Christian and said, "Thank you so much!"

Kennedy said, "I can't wait to show my mom."

As the girls continued to fly their kites high in the sky, they giggled as they heard the sounds of their voices.

The four friends laughed and watched as the girls flew their decorative kites for several minutes.

___________ remarked, "I can't believe you and your dad added glitter to each kite."

"It took me three hours to wash off the glitter before I went to bed. But seeing their smiles is ALMOST worth it! I dislike glitter," Christian said with a half-smile.

Sarah said, "I could just sit here and watch them all day."

_____________ said, "Me too!"

After 30 minutes, Kennedy, Vanessa, and Mary pulled down their kites and sat with the four friends on a big blue blanket.

Vanessa and Mary said, "Thank you, Cousin Christian."

Christian smiled and said, "You're welcome. I'm so glad you're having fun with your kites. Vanessa, could I show you a few pictures?"

Vanessa said, "Do I get to color them?"

Christian said, "I'm sorry. I didn't bring crayons. Next time I'll bring coloring books and crayons." Christian pulled out four photographs and asked, "Vanessa, have you ever seen these pictures and books?"

Vanessa looked at the photographs and said, "Yeah, my brother Pap gave them to me. Math makes me happy. I love Pap."

The friends looked stunned and then looked at each other. No one said a word.

Just then, Pap walked over to the group.

Pap said, "Hi, everyone. Are you flying kites today?"

The four friends looked at Pap.

Vanessa said, "Look, Pap. Christian made us kites."

Pap said, "Thank you, Christian. Butterflies, kites, and math make Vanessa happy."

Vanessa said, "Pap, Christian has pictures of the gifts you gave me."

Pap saw the photos. He looked at Vanessa and then looked at the four friends.

Everyone was quiet.

Sarah said, "Kennedy, Vanessa, and Mary, let's walk over to the ice cream stand. I'll buy your favorite flavors."

Kennedy said, "We LOVE ice cream!" The three girls followed Sarah to the ice cream stand.

Pap was quiet and then asked, "Christian, could I speak to you alone?"

The other two friends stood up and walked over to the ice cream stand.

Pap asked Christian, "How did you know?"

Christian whispered, "Sarah and I have special dreams, and they lead us to you and Vanessa. And when I read the poem you mailed me. I was sure it was you. You told us your sister loves math and butterflies. It must be tough to have someone you love struggle with a disability."

"It's difficult, and I was sure you'd discover I took the objects. It was my way of explaining it. Vanessa has been so sad. We lost our dad last year. My mom received some news at the end of the school year from Vanessa's doctor. She has a health condition, and the news broke our hearts. I love her so much and would do anything to make her happy. It was wrong, and I'll tell my mom and go to the police. But each time I gave her an object and explained its connection to math, she'd giggle, clap, and twirl."

Sarah walked over to Pap and Christian while their two friends sat on another side of the park with the girls and Kennedy's mom and ate their ice cream.

Sarah asked Pap, "I've never seen Vanessa's bedroom. How's it decorated?"

Pap asked, "Why?"

Sarah said, "I think Christian and I had dreams of Vanessa and her bedroom."

"Vanessa's room is drenched in butterflies. There are seven butterflies painted on her ceiling. She has a large bookcase filled with books, and I added the two books I took next to her book on *Symmetry*. She loves to wear her butterfly dress, with her butterfly sneakers and butterfly hair clip, and twirl around and dance in her bedroom. I always pretend I'm scooping the stars and numbers out of the sky with her small butterfly net. Then I drop them into her hands."

Sarah and Christian looked at each other and smiled. Sarah said, "Well, it makes sense why the butterflies in our dreams couldn't fly very high. The butterflies are painted on the ceiling in Vanessa's bedroom."

Christian said, "Pap, I want your permission to call my Uncle Raymond. He's the police detective responsible for this case. Sarah and I'll meet you at the police station tomorrow at 1:00. We want to be there with you. Please bring all the objects, and you should tell your mom tonight."

Pap said, "I will. Why aren't you calling the police, and why are you both going with me?"

Sarah said, "*The Golden Rule*," and smiled and said, "*Not the Golden* ______. 'Do unto others as you want them to do unto you.'"

Christian added, "I hope everyone will understand, and while you must give everything back, I hope you won't be in trouble. We'll do everything we can to help you out. We understand you just wanted to make your sister happy. I'm sorry she has health problems. She's so sweet and so smart. You must be proud of her."

Pap said, "I am, and thank you. I'll meet you both at the police station tomorrow at 1:00 and bring the objects."

The friends walked home, and Christian explained everything. He asked his friends not to talk about what happened or what he told them.

◆ ◆ ◆

When Christian arrived home, he called his Uncle Raymond, who drove to Christian's house. Christian explained everything.

Uncle Raymond said, "Well, he's returning everything. I understand why he took the objects. If he hasn't damaged any of the items, I'll see what I can do. Christian, you four solved another mystery hunt! I hope this will be a happy ending. I'll contact the owners of the objects, and speak to my supervisor. He's not as understanding as I would like. See you and Sarah tomorrow with Pap at 1:00."

"Thank you, Uncle Raymond. He's such a nice guy, and he loves his sister. He just graduated from college and is looking for a job. I don't want this to hurt his life. We all feel the same way."

Uncle Raymond said, "I know, Christian. I'll see what I can do."

◆ ◆ ◆

The next day Pap arrived at the police station. He could see Christian and Sarah sitting by a wall outside of an office. Two men were in the office having, what looked like, a heated discussion.

Pap walked over to his two friends.

Christian said, "Have a seat. My uncle is speaking to his boss about the case. I'm sure my uncle will do what he can for you."

Detective Dawson's boss walked out of the office, slammed the door, and gave a stern look at the three friends sitting along the wall. He stomped off down the hallway.

Uncle Raymond opened his door and waved the three in.

"Hello, Detective Dawson. My name is Pap Nova. I know Christian explained I'd be here today. Christian and Sarah, thanks for being here with me."

Detective Dawson shook Pap's hand. "Pap, please follow me. Sarah and Christian, you can come with us. Pap, please bring the box with the objects."

Pap's hands were shaking. Sarah grabbed his hand.

Uncle Raymond opened the door to a large room. Six people were sitting in the room. Detective Dawson introduced Pap to Ms. Parvana, Mr. Bigollo, Ms. Draco, Miss Bell, and Mr. Apar. The Science Museum Director, Dr. Theis, stood up and shook Pap's hand.

Detective Dawson said, "Pap, what you did was wrong, but it was for the love of your sister. I've spoken with everyone, and no one will press charges. Are the objects intact?"

"Yes, sir." Pap had a few tears in his eyes and said, "I need to apologize. What I did was wrong. I love my sister so much. She's the kindest and most loving person I know. She was so sad before I returned home. Math objects and butterflies always make her smile. I explained everything to my mom. She wanted to be with me today, but I told her I had to do this alone. I apologized for letting her down, and again, please accept my apologies for any worry I caused each of you. My friends, Christian and Sarah, are so kind and supportive, and I'm grateful. I'm searching for a job, and I'm happy to pay any compensation to each of you."

Ms. Parvana said, "Pap, we forgive you. We all make mistakes. It's part of being human. I love Vanessa. I don't have children of my own, so please keep the framed butterfly for Vanessa. She brings me joy, and I hope the butterfly will bring her joy."

Mr. Bigollo stood up and said, "Thank you for returning our books and objects. We forgive you. I brought a box filled with mathematical children's books my grandfather gave to me and other objects. Each of us donated a meaningful math object for your sister Vanessa. A friend of mine owns a small research lab. Detective Dawson informed me you completed college with a degree in Mechanical Engineering. Here's his business card. Stop by his office tomorrow to complete an application. He has two entry-level positions available."

Pap had tears in his eyes. "Thank you for your forgiveness. I'll never forget each of you and the life lesson on *The Golden Rule*. My sister and my family will be so grateful for your forgiveness and your kindness. Thank you!"

Christian stood up and said, "May Sarah and I read something attributed to Einstein? I'm not sure he wrote it, but Sarah found this online."

Everyone nodded and smiled, and Sarah began reading the quotation.

There is an extremely powerful force that, so far, science has not found a formal explanation to. It is a force that includes and governs all others and is even behind any phenomenon operating in the universe and has not yet been identified by us. This universal force is LOVE. When scientists looked for a unified theory of the universe, they forgot the most powerful unseen force. Love is light that enlightens those who give and receive it. Love is gravity because it makes some people feel attracted to others. Love is power, because it multiplies the best we have, and allows humanity not to be extinguished in their blind selfishness. Love unfolds and reveals. For love, we live and die.

Christian continued reading the quote.

Love is God and God is Love. This force explains everything and gives meaning to life. This is the variable that we have ignored for too long, maybe because we are afraid of love because it is the only energy in the universe that man has not learned to drive at will. To give visibility to love (I/we) made a simple substitution in (my) (Einstein's) most famous equation. If instead of E = mc2, we accept that the energy to heal the world can be obtained through love multiplied by the speed of light squared, we arrive at the conclusion that love is the most powerful force there is, because it has no limits. If we want our species to survive, if we are to find meaning in life, if we want to save the world and every sentient being that inhabits it, love is the one and only answer.

Everyone smiled. Ms. Parvana and a few of the others had tears in their eyes. There was absolute silence in the room.

Christian and Sarah helped Pap out with his box of 'goodies' for his sister Vanessa.

Pap said, "You're amazing Mystery Hunters and even better friends. Thank you. I'll never forget you."

Sarah said, "Give Vanessa a hug and kiss for us, and always remember to make her giggle and continue to drop numbers and stars in her hands."

Uncle Raymond walked out and waved to Pap as he drove away. He said, "So everything turned out all right. I wanted to let the two of you know I checked into Asst. Director Apar's vehicle accident. Not only did the two drivers report seeing a shadow, but a man about to cross the street told the police he saw something strange. The service garage reported that the brakes on both cars were working fine."

Sarah said, "Thanks. I'm glad no one was hurt."

Christian asked, "Are you in trouble with your new boss?"

"Well, it's nothing I can't handle. My boss insisted on Pap being arrested and wanted to call the media to give a press conference. I stood my ground, but he isn't happy with me." He chuckled and said, "Maybe my boss is the 'invisible force.'" He hugged Christian and Sarah and walked back into the police station.

Christian and Sarah walked home.

Sarah said, "Remember your dream of the door with bars being slammed shut? Uncle Raymond's boss wanted Pap to be arrested, but the invisible force didn't win. Everyone forgave Pap. Good wins over evil!"

Christian said, "This invisible force is starting to worry me. I don't want you or anyone else to get hurt. When Tata said we'd have to discover who or what this force is, well ... I'm sure it's an evil force. It probably has its hands on many things, but maybe it focuses on us because of our prophetic dreams. It's not just targeting us, but anyone we try to help."

Sarah said, "I know some people don't believe in evil. But if there's good, there's evil. I think we have to be more aware of our surroundings and how we handle any new cases. Even though I'm a little scared, I don't want to stop helping people."

"I agree."

Sarah smiled and said, "On a lighter note, I believe the three little girls in your dream were Vanessa, Mary, and the third little girl was Einstein's daughter, Lieserl.

She was encouraging you to show kindness and love to Vanessa and Mary, and in her special way, to Pap."

Sarah and Christian looked at each other and said, "No coincidences." Little did they know their next adventure would bring new dreams ... transporting them around the corner and through the universe and back.

SYMMETRY IN FLIGHT

Authors Bio: Ramona Palmer and her son Kurtis were inspired to write a series of children's books. She and Kurt, her husband of over thirty years, read to their son before his birth and beyond. They believe this instilled a love for books and learning in their son. By the age of three, Kurtis asked for books for every birthday and Christmas. After several years his mom, tired of dusting hundreds of books, suggested he build a couple of libraries in his own home someday. Ramona and Kurtis both have degrees in Communications, with Kurtis completing a 3D animation program. The family lives in Upstate New York.

Their family advocates reading to a child in utero and beyond, naming their company, ***Womb to BLOOM to Classroom***. We now know that prenatal babies can hear and react to words and sounds and that 90% of a child's brain develops by the age of 5! The family believes, and the research supports, reading to and with children fosters imagination while entertaining, educating, elevating, and inspiring them to love reading and learning.

The Mystery Hunters Series begins with ***The Lost Polymath***. Sarah and Christian have mysterious *dreams* that provide clues to help solve each mystery on their journey, while an unseen force tries to stop them from completing their quest. This series engages the reader and a friend to become members of the team.

The first picture book in their ***Womb to BLOOM to Classroom*** series, ***Numbers from Heaven***, begins with the story of Zoey Zebra and her best friend P.B., the panda bear. They discover a treasure chest left for Zoey by her grandpa that takes them on treasure hunts. This series has over 50 beautifully illustrated books by artist Kimmo Hellström on Mathematics, Science, Zoology, Literature, and more.

From their family to yours, enjoy reading to and with the children you love and care for each day. ***Visit us at: www.SmarterBrighterKids.com****!*

Made in the USA
Monee, IL
08 November 2026

67388657R00057